The Pocket Guide to
Astronomy

Patrick Moore

Simon and Schuster
New York

Contents

The Pocket Guide to Astronomy
was edited and designed by Mitchell Beazley Publishers
14–15 Manette Street, London W1V 5LB
©1980 by Mitchell Beazley Publishers Limited
Revised edition 1982
Updated and reprinted 1985
All rights reserved
Published by Simon and Schuster
A Division of Simon & Schuster Inc.
Simon & Schuster Building,
Rockefeller Center,
1230 Avenue of the Americas,
New York, New York 10020
Moore, Patrick.
    The pocket guide to astronomy.

    1. Astronomy. I. Title.
QB64.M63   520   79-5382
ISBN 0-671-25309-3
Printed and bound in Hong Kong

Preface

This pocket guide is a working manual for serious amateur observers all over the world. It is divided into four sections: the opening one, Introduction to Astronomy, covers astronomical terminology and observational techniques. The two major sections, the Solar System and the Stars, provide a detailed guide to the heavens. They incorporate not only the latest information from the space probes, in colour plates and data, but also all the background essential to understanding the stars and planets. The final section contains detailed appendices, including indexes to the Moon maps and seasonal star charts. In addition there are tables giving the locations, over the next four years, of the observable planets, the dates of lunar and solar eclipses and comet and meteor shower returns.

The seasonal star charts are backed by individual location maps of the most interesting celestial objects, orientated to reflect the actual views seen by the observer in either hemisphere. The Moon maps are orientated with south at the top to suit Northern Hemisphere observers using telescopes. The descriptive comments accompanying the maps and charts are based on perfect viewing conditions.

Finally, in keeping with current scientific practice, all measurements are metric. For those more familiar with English measurements, the following table will be of use.

METRIC CONVERSION TABLE					
in to cm		cm to in	miles to km		km to miles
2.54	1	0.39	1.61	1	0.62
5.08	2	0.79	3.22	2	1.24
7.62	3	1.18	4.83	3	1.86
10.16	4	1.58	6.44	4	2.49
12.70	5	1.97	8.05	5	3.11
15.24	6	2.36	9.66	6	3.73
17.78	7	2.76	11.27	7	4.35
20.32	8	3.15	12.88	8	4.97
22.86	9	3.54	14.48	9	5.59
25.40	10	3.94	16.09	10	6.21
50.80	20	7.87	32.19	20	12.43
76.20	30	11.81	48.28	30	18.64
101.6	40	15.75	64.37	40	24.86
127.0	50	19.69	80.47	50	31.07
152.4	60	23.62	95.56	60	37.28
177.8	70	27.56	112.7	70	43.50
203.2	80	31.50	128.7	80	49.71
228.6	90	35.43	144.8	90	55.92
254.0	100	39.37	160.9	100	62.14

Understanding the night sky 1

The Earth is a planet, 12,765 kilometres in diameter, which moves round the Sun once every 365.86 days at a mean distance of 150 million kilometres. It is a member of the Sun's family or Solar System and there are eight other planets in the system, some of which have satellites orbiting them. The Earth has one satellite, the familiar Moon. The stars themselves are suns, and these are so remote that even their light, moving at 300,000 kilometres per second, takes years to reach the Earth.

Because the stars are so far away, their individual or "proper" motions are very slight and the patterns or constellations now known to astronomers are the same as those that must have been seen by prehistoric man. It is convenient to picture the Earth as being surrounded by a sphere, the celestial sphere, which has a centre the same as the centre of the Earth. The various objects in the sky may then be positioned on this imaginary sphere.

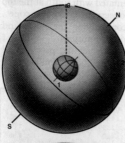

The celestial poles (N and S) are the points on the celestial sphere indicated by the direction of the Earth's axis; the north celestial pole is marked to within 1° by the bright star Polaris; there is no equivalent star to mark the south celestial pole. The celestial equator (1) is the projection of the Earth's equator on to the celestial sphere. The meridian (2) is the great circle on the celestial sphere that passes through both poles and zenith (3).

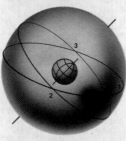

The ecliptic is the apparent yearly path of the Sun among the stars. Its inclination (1) to the celestial equator is the same as that of the Earth's axis to the orbital plane: thus the Sun stays six months in the Northern Hemisphere and six months in the Southern. The points where the Sun crosses the equator are the equinoxes: the vernal equinox (about 21 March) or the first point of Aries (2) and the autumnal (about 22 September) or the first point of Libra (3).

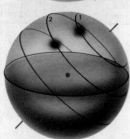

The altitude of the celestial pole above the horizon is equal to the observer's latitude. From the Earth's pole the celestial pole will be at the zenith—altitude 90°. From an observer at latitude 40°N the north celestial pole will be at altitude 40°. Stars close to the pole (1), such as those in Ursa Major seen from central Europe and the northern United States, are circumpolar and will never set. Stars further from the pole (2) will fall below the horizon.

In the sky the equivalent of latitude on Earth is known as declination: that is, the angular distance north or south of the celestial equator. The celestial equivalent of longitude is known as right ascension, and this is the angular distance of a body from the first point of Aries, measured eastwards; it is usually measured in units of time rather than in degrees. As the Earth rotates, a star will seem to rise, reach its highest point or "culmination" and then set. The first point of Aries must therefore culminate once a day. The right ascension of an object is the time interval between the culmination of the first point and the culmination of the object concerned.

Apart from the effects of "precession", a minor shift in position of the celestial poles and equator due to a slight "wobbling" of the Earth's axis, the right ascensions and declinations of the stars do not change appreciably. Thus a star's celestial coordinates can be sufficiently well established using the celestial equator as the reference plane.

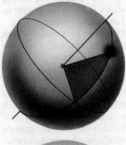

Latitude, on the Earth, is defined as the angular distance from the centre of the Earth, north or south of the equator. For celestial bodies the corresponding term is declination (1). Thus the north celestial pole has declination N 90°; the declination of the celestial equator is zero. The declinations of all celestial bodies take values between 0° and 90°, described as positive for an object north of the celestial equator and negative for an object south.

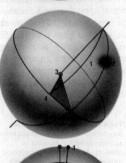

The prime meridian (1) is the great circle running north-south on the celestial sphere through the first point of Aries. The angle between the prime meridian and a celestial body is known as right ascension. The right ascension of a star (2) is the angle from an observer (3) between the first point of Aries (4) and the point on the celestial equator vertically below the star (5). Declination and right ascension define a star's position.

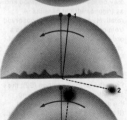

Right ascension is measured in time units where 24 hours equals 360° When the first point of Aries (1) culminates the time is 00 h 00 m. The star Sirius (2), which is then below the horizon, culminates 6 h 44 m later, having RA 6 h 44 m. This apparent 360° rotation of the celestial sphere is called the sidereal day. It is, however, about 4 minutes shorter than the solar day because the Sun, which completes the 360° in 365 days, moves eastwards by about 4 minutes a day.

Understanding the night sky 2

Astronomers are used to vast distances and immense spans of time. The Moon at a mean distance of 384,400 kilometres from the Earth is its closest neighbour; even the Sun (150 million kilometres away) is relatively very near the Earth. When astronomers measure the distances of the stars, ordinary units become hopelessly inconvenient. Instead, astronomers use the term "light-year", which is the distance travelled by light in a year. Light moves at 300,000 kilometres a second, so one light-year is equal to 9.46 million million kilometres. Even the nearest star beyond the Sun is greater than four light-years away; this is why the "proper" motions of the stars are so slight. The star with the greatest known proper motion (Barnard's Star, a faint red dwarf) takes 180 years to crawl across the sky by an amount equal to the apparent diameter of the full Moon.

A "constellation" has no real meaning as the stars are not at equal distances from the Earth; it is simply a pattern of stars that happen to lie in much the same direction, as seen from Earth. From a different vantage point in the universe, the constellation patterns would look quite different.

The stars are suns, but there is a tremendous range in both size and luminosity. The apparent magnitude of a star is a measure of its brightness as seen from Earth. The scale works like a golfer's handicap, with the more brilliant performers having the lower values. Thus bright stars are of magnitude 1 or even less (the four brightest stars in the sky have negative magnitudes), while stars of magnitude 6 represent the limit of naked-eye visibility under good conditions and large telescopes can reach well below magnitude +20. Everything, however, depends upon the star's distance from Earth: for example, Vega (magnitude 0.04) is much brighter than Deneb (magnitude 1.3), yet Vega is a mere 55 times as luminous as the Sun, while Deneb is equal to perhaps 60,000 Suns. This means that Deneb must be the more remote of the two and that its absolute magnitude is far greater (absolute magnitude being defined as the apparent magnitude that a star would have if it were located at a standard distance from Earth).

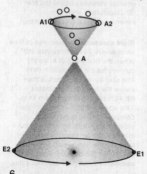

The distances of the nearer stars may be measured by a method known as trigonometrical parallax. A relatively close star (A) is observed from opposite sides of the Earth's orbit, in other words at an interval of six months. From the Earth's first position (E1), the star will be seen in position A1 against the background of more remote stars. From the second position (E2), the star will appear at A2. Knowing the diameter of the Earth's orbit (E1–E2) and knowing the apparent angular shift, the observer will then be able to calculate the distance of the star from Earth.

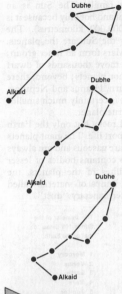

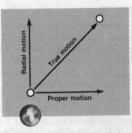

Actual motions over a long period will alter the constellation patterns as seen from Earth. True motion of a star is a combination of proper or ''sideways'' motion and radial or ''towards or away'' motion (*above*). In the constellation of Ursa Major, consisting of seven distinct stars, two stars (Alkaid and Dubhe) are more remote than the rest and are moving in different directions, so that eventually the pattern will be distorted (*left*). Distances (*below*) range from 63 light-years (Megrez) to 210 (Alkaid), so that Alkaid is much farther away from Megrez than Megrez from Earth.

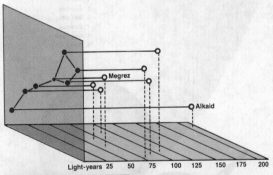

Megrez

Alkaid

Light-years   25   50   75   100   125   150   175   200

The Crab Nebula is the result of a violent stellar explosion or supernova. This particular outburst was seen from Earth in 1054 but, because the Crab Nebula is 6,000 light-years away, the actual outburst occurred some 6,000 years earlier—before there were any astronomers to observe it. At its brightest, the 1054 outburst would have been visible with the naked eye in broad daylight. Since then astronomers have observed only two supernovae within the Galaxy—these were stars that were recognized in 1572 by Tycho Brahe and in 1604 by Johannes Kepler.

The Solar System 1

The Solar System is the Sun's family. The Sun is an ordinary star; it appears so brilliant and hot only because it is relatively close to Earth (150,000,000 kilometres). The principal members controlled by the Sun are the planets. Mercury, Venus, the Earth and Mars form an inner group; then comes a wide gap, in which move thousands of dwarf worlds known as asteroids or minor planets; beyond, there are the giant planets Jupiter, Saturn, Uranus and Neptune. Finally there is Pluto, a body that is certainly much smaller than the Earth and may not be a true planet.

The inner planets are solid and rocky but only the Earth has an atmosphere suitable to support life. The giant planets are quite different; their surfaces are gaseous and are always changing. The Solar System also contains bodies of lesser importance: the moons or satellites of the planets, the flimsy and insubstantial comets, lumps of material called meteoroids and a great deal of interplanetary "dust".

The diameter of the
Sun is 109 times
that of the Earth.

1 Mercury
2 Venus
3 Earth
4 Mars
5 Jupiter
6 Saturn
7 Uranus
8 Neptune
9 Pluto

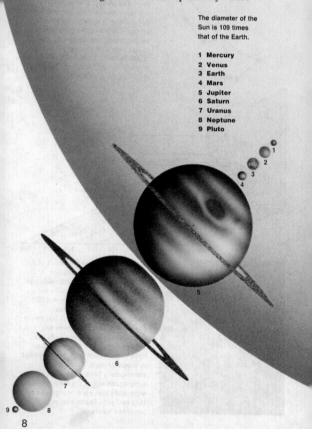

Mercury	Venus	Earth	Mars	Jupiter	Saturn	Uranus	Neptune	Pluto	
69.7	109	152.1	249.1	815.7	1,507	3,004	4,537	7,375	Max. distance from Sun (millions of km)
45.9	107.4	147.1	206.7	740.9	1,347	2,735	4,456	4,425	Min. distance from Sun (millions of km)
57.9	108.2	149.6	227.9	778.3	1,427	2,869.6	4,496.6	5,900	Mean distance from Sun (millions of km)
0.39	0.72	1.0	1.52	5.20	9.54	19.18	30.06	39.44	Mean distance from Sun (astronomical units)
0.24	0.61	1.0	1.88	11.86	29.46	84.01	164.8	247.7	Period of revolution (years)
47.9	35	29.8	24.1	13.1	9.7	6.8	5.4	4.7	Mean orbital velocity (km per second)
0.0°(?)	2°	23°27'	24°46'	3°05'	26°44'	97°53'	28°48'	?	Inclination of axis
7°	3.4°	0°	1.9°	1.3°	2.5°	0.8°	1.8°	17.2°	Inclination of orbit to ecliptic
0.206	0.007	0.017	0.093	0.048	0.056(?)	0.047	0.009	0.25	Eccentricity of orbit
4,880	12,104	12,756	6,787	142,800	120,000	51,800	49,500	3,000(?)	Equatorial diameter (km)
0.38	0.90	1.0	0.38	2.64	1.16	1.11	1.21	?	Surface gravity (Earth=1)
None	CO_2	N_2O	CO_2/Ar	H.He	H.He	H.He.CH_4	H.He.CH_4	CH_4 ?	Atmosphere (main components)
350(S) day −170(S) night	−33(C) day 480(S) night 22(S)	−23(S)	−150(C)	−180(C)	−210(C)	−220(C)	−230(C)		Mean temperature at surface (degrees Celsius) S=solid C=clouds
59	−243 retrograde	1.0	1.03	0.41	0.43	−0.6? retrograde	−0.77	6.38(?)	Rotation period (days)
0	0	1	2	16	21	5	2	1	Known satellites
0.055	0.815	1	0.108	317.9	95.2	14.6	17.2	0.0025	Mass (Earth=1)
0.06	0.88	1	0.15	1,316	755	67	57	0.015 (?)	Volume (Earth=1)
5.5	5.3	5.5	3.9	1.3	0.7	1.7	1.8	2(?)	Density (water=1)
0	0	0.003	0.009	0.06	0.1	0.06	0.02	?	Oblateness
Mercury	Venus	Earth	Mars	Jupiter	Saturn	Uranus	Neptune	Pluto	

The Solar System 2

The planets were formed from a "solar nebula", a cloud of material associated with the youthful Sun. They all move around the Sun in the same direction. There is no example of a planet with a retrograde or "wrong-way" motion although some comets are retrograde, as are some satellites of the planets. Moveover, the orbits of most planets are in roughly the same plane, with inclinations of less than 4 degrees. The two exceptions are Pluto, with an inclination of 17 degrees and Mercury, 7 degrees.

Mercury and Venus, closer to the Sun than Earth, are known as the "inferior" planets and show phases like those of the Moon. Mars and the outer or "superior" planets are more convenient to observe. When a superior planet is at opposition the Sun, the Earth and the planet are in a straight line and well placed for study.

The revolution periods—the time that it takes for a planet to complete one revolution around the Sun—range from 88 days for Mercury to almost 248 years for Pluto. The orbits take the form of wide-angled ellipses; drawn on a small scale they appear as circles. Again Pluto is exceptional: its orbit is much more eccentric and when at perihelion, or closest point of approach to the Sun, it comes within the orbit of Neptune. It is next due at perihelion in 1989, so that at present Neptune, not Pluto, ranks as "the outermost planet" in the Solar System.

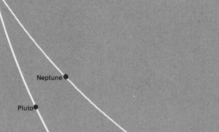

Uranus ●

Neptune ●

Pluto ●

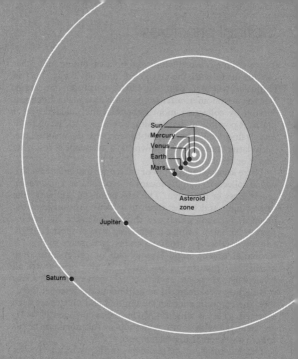

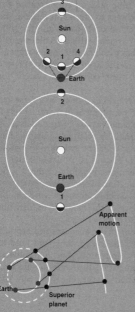

The conjunctions and elongations
of an inferior planet are shown
(*left*). At position 1 the planet is new
(inferior conjunction); at 2 it is at
elongation, or maximum apparent
distance from the Sun; at 3, full
(superior conjunction) and at 4 it
has returned to elongation. If
the alignment at inferior conjunction
is perfect, the planet appears in
transit against the Sun's disk.
Mercury will next transit in 1986,
Venus not until 2004. The
opposition and conjunction of a
superior planet are shown (*centre*).
Opposition occurs when the planet
is aligned with the Sun and Earth
at position 1 and is then best
placed for observation. At superior
conjunction (2) the planet is on the far
side of the Sun and is unobservable.

**The apparent retrograde movement
of a superior planet** is shown (*left*).
In general the planet moves from
west to east against the starry
background but, when the Earth
is about to "overtake" it, the planet
seems to stop, move "backwards"
for a while, stop once more, and
then resume its eastward motion.
Mercury and Venus also do this.

11

Optical principles

There are many worthwhile astronomical observations that can be made with the naked eye but obviously the real enthusiast will soon want to graduate to using an optical aid. For the astronomer, a telescope fulfils two main functions. Firstly, by collecting more light than the unaided eye, it makes it possible to detect fainter objects; and, secondly, it reveals fine details that cannot be seen with the unaided eye. This latter function is known as a telescope's resolving power or resolution.

There are two main types of telescope: first, the refractor, which collects its light by means of a special lens known as an object glass or objective. The lens bends or "refracts" the rays of light and brings them to a focus. An image of the object is formed, which can then be enlarged by an eyepiece—this latter being a magnifying glass of special design. Initially a telescope produces an inverted image so that for terrestrial use an extra lens system or prism is used to correct the picture. (This feature is also present on all binoculars.) However, each time light passes through a lens it is slightly weakened. In an astronomical telescope, therefore, the correcting lenses are omitted and images are seen inverted. (This is why the Moon maps, on pages 26 to 41, are arranged with south at the top to suit northern hemisphere observers using telescopes.)

The second kind of telescope is the reflector. In the Newtonian pattern (so called after its inventor Sir Isaac Newton), the rays of light strike a curved mirror and are reflected back up the open tube and on to a smaller flat mirror, which directs the light rays to the eyepiece set in the side of the tube. Therefore the observer looks into the tube and not up it. In an alternative system, which is known as the Cassegrain, the main mirror has a hole in its centre so that all light rays are reflected back to the eyepiece. However, the Newtonian is probably the most convenient system for the amateur to use.

It is unwise to spend a great deal of money on a refractor with an aperture less than 7.5 centimetres in diameter or on a reflector with a main mirror smaller than 15 centimetres. A very small telescope, which may be relatively inexpensive, will have a small field of view and will be awkward to handle as well as being low powered. Thus it is probably better to invest in a good pair of binoculars, possibly of the 7×50 variety (that is, with a magnification of seven, with each lens 50 millimetres in diameter). Such binoculars will not be capable of showing, for example, the rings of Saturn but they will certainly be an aid to observation.

Refractors are more efficient than reflectors and are less liable to damage but they are more expensive and tend to give a certain amount of false colour around brilliant objects such as stars. The amateur can grind a mirror for a reflecting telescope himself: this method is the cheapest but the difficulties are considerable.

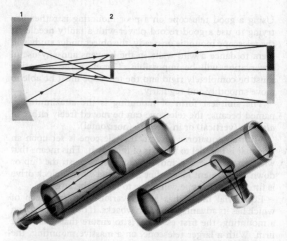

In a reflector (*top*) the light is collected by a concave primary mirror (1) and brought to focus, forming an image (2). The size of the image depends directly on the focal length of the instrument—the distance between the mirror and the image. In a Cassegrain

reflector (*above left*) light from the primary mirror is intercepted by a convex secondary and reflected back through a hole in the primary. In the Newtonian reflector (*above right*) light is reflected from the primary on to a flat secondary set at 45° to the telescope's axis.

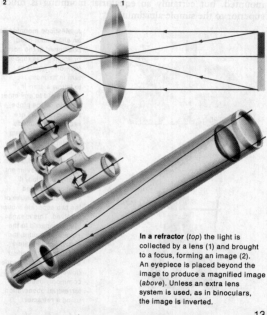

In a refractor (*top*) the light is collected by a lens (1) and brought to a focus, forming an image (2). An eyepiece is placed beyond the image to produce a magnified image (*above*). Unless an extra lens system is used, as in binoculars, the image is inverted.

13

Telescope mountings

Using a good telescope on a poor mounting is rather like trying to use a good record player with a faulty needle. If the telescope mount is unsteady, the object under study will seem to dance a wild waltz in the heavens and any useful observation will be impossible. Therefore the mounting must be completely rigid and the telescope must be able to move smoothly and regularly.

The simplest form of mounting is the altazimuth, so named because the telescope can be moved freely either in *alt*itude (vertical) or in *azimuth* (horizontal).

With an equatorial mount, the telescope is set upon an axis that is parallel to the axis of the Earth. This means that when the telescope is moved from east to west, the "up or down" movement looks after itself; and when a clock drive is fitted the telescope will move to follow the target.

Equatorial mountings are of various patterns, each of which has its advantages and drawbacks. If buying or making a mounting, the first essential is to ensure that it is really firm. With a larger telescope, on a massive mounting, the installation must be permanent. Generally speaking a Newtonian reflector with an aperture of 20 centimetres (that is, with a mirror 20 centimetres across) is on the limit of portability, even if the tube is a skeleton construction instead of being made of sheet metal or something equivalent.

It would be wrong to claim that all telescopes powerful enough to show significant detail must be equatorially mounted, but certainly an equatorial mounting is much superior to the simple altazimuth.

Equator

A telescope mounted on an altazimuth head, positioned at the north or south poles (*upper left*), only needs to turn in azimuth to follow a star. If it is moved to any other latitude on the globe two movements are required, in altitude and azimuth, to keep the stars in view. An altazimuth mount is usually a tripod, which allows free movement in any direction. It cannot be guided very easily because of the two separate motions involved. This means that, compared to the equatorial mount, the altazimuth is mainly suited to the smallest astronomical telescopes (*lower left*). It is more commonly used for terrestrial observation, using a refractor.

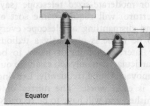

Equator

A telescope mounted on an equatorial head always has its polar axis pointing at the celestial pole, wherever it is positioned on the globe. When the telescope is moved from east to west, it automatically compensates for the changing altitude.

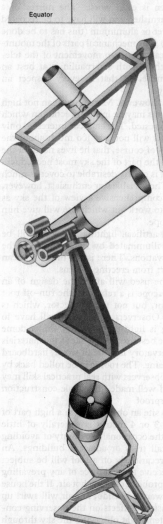

The Yoke design has two pillars carrying an axis on which the telescope is mounted. Its main advantage is that it does not require a counterweight. However, the polar region is permanently inaccessible and so, to a certain extent, is the area below it at any time. Also the mounting is unsuitable for refractors and Cassegrains.

The German mount is particularly suited to refractors. The telescope is mounted on one end of the axis with a counterweight at the other. It can be used in all latitudes; it is relatively simple to construct, accessories are easily added and the whole sky can be observed. The main disadvantage is the necessity for a counterbalance, which increases the weight.

The Fork mounting is the most suitable for reflectors. There is no cumbersome counterweight and the design is particularly stable. If the polar axis and fork are made sufficiently rigid, the mounting is probably the best that can be used in conjunction with the Newtonian reflector—one of the more popular amateur telescopes; it is used in many of the world's largest observatories.

15

Home observatories

The owner of a large- or moderate-sized telescope (say, above 24-centimetre aperture) will require some sort of observatory. Erecting and dismantling a telescope every time observations are to be made is obviously a tedious business and, sooner or later, something is likely to be dropped. Reflectors, in particular, can only too easily go out of adjustment if they are moved around frequently. Leaving a telescope in the open is even worse. The mirror of a reflector will become tarnished and will have to be recoated with a thin layer of silver or aluminium (this has to be done periodically in any case). The mechanical parts of the mounting will rust and may cause uneven movement of the telescope. Covering everything with a tarpaulin is at best an unsatisfactory compromise, so that in every respect an observatory is necessary.

There are difficulties, however. More often than not high trees or other obstructions may lie in the direction in which the observer is most interested. The amateur interested only in the Moon and planets will not care too much about the northern sky (assuming, of course, that he lives north of the Earth's equator) and so the part of the sky most needed will be the south although it is clearly desirable to cover as much of the sky as possible. The variable star enthusiast, however, will ideally require as comprehensive a view of the sky as possible and will have to work out which site will give him the most reasonable scope.

Another nuisance is artificial lighting. Viewing will be hindered if the sky is illuminated by street lamps in the direction of your observations. There is not much that can be done about this apart from erecting screens.

The type of telescope used will affect the design of an observatory. If the telescope is a refractor, the run-off roof arrangement is excellent, but not for a reflector, which is mounted lower down. Observers of the Sun will have to ensure that everything is light-tight and for this a dome structure is certainly the best method. So far as the materials are concerned the observatory may be of wood, hardboard or plastic with a firm frame. The roof may be rolled back by a pulley arrangement; observers with real practical skill may prefer to motorize it. If well made, the whole construction should be fully weatherproof.

It is not necessary to site an observatory in a high part of the garden. Going up 3 or 4 metres is generally of little advantage, apart from the occasional possibility of avoiding obstructions such as tall trees or adjacent buildings. An observatory perched precariously on a roof will be subject to vibration and will receive the full force of any prevailing wind. Worse still is the problem of rising hot air. If the house is inhabited, it will be warmed inside; hot air will swirl up over the roof with disastrous effects on the observing conditions. This is why attempts at observing the sky through telescopes at bedroom windows are doomed to failure.

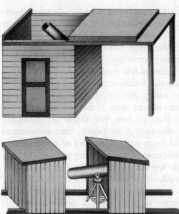

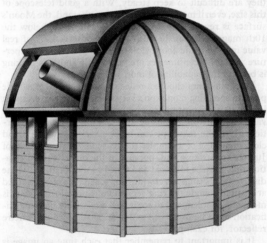

The run-off roof is designed to slide back out of the way, leaving a shelter for the observer. The design is best suited for observers of the Sun; with a reflector, the view of the sky is restricted. The run-off shed is easy to make and use; however, the viewer has to work in the open and endure both cold and wind as well as scattered light. The best design consists of two parts (*left*); the alternative single shed with a door is more cumbersome and not as portable when changing the site.

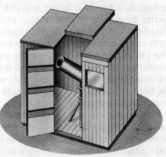

The dome observatory (*above*) is difficult to make. The upper part, the hemispherical dome itself, moves round so that the slit can be directed to any point in the sky. Part of the problem of construction is making the rail, on which the dome rotates, absolutely circular. The same problem arises with the total rotator (*right*); here the whole observatory moves round a circular rail on wheels.

Making observations

It is often thought that no useful or even satisfactory observations can be made without the help of a powerful and very expensive telescope. This is not so. Of course, the larger the telescope the greater the range but things are not nearly so daunting as the beginner sometimes believes. For instance, binoculars will give splendid low-power views of the Moon's surface as well as rich star fields, clusters and nebulae. The phases of Venus are easily discernible and the four large satellites of Jupiter can also be seen. If the binoculars have a magnification higher than 12, however, the field is so small, and the binoculars themselves so heavy, that it becomes difficult to hand-hold them without the aid of some kind of rest. Secondhand telescopes of good quality are rare for they may have defects that are not obvious at first glance.

Small refractors with object glasses from 2.5 to 6.5 centimetres across are of limited value. The minimum useful aperture for a refractor is probably 7.5 centimetres but, although refractors will show more detail than binoculars, they are difficult to keep steady. With a good telescope of this size, even if mounted on an altazimuth stand, the Moon's surface is revealed in considerable detail. Stars below the 10th magnitude can be seen so that observations of real value may be made about variable stars. Refractors of aperture above 7.5 centimetres are very costly and lens-making is beyond the capability of most people.

Most amateurs choose a reflector and again there is little point in spending money on a very small one, that is one with less than a 15-centimetre aperture. Many amateurs make valuable contributions using a 15-centimetre reflector, particularly if the telescope can be equatorially mounted and clock driven. For example, the features on the surface of Jupiter are striking and, because of the planet's rapid rotation they may be followed and timed as they drift across the disk. Saturn's rings are glorious and when the system is tilted at a suitable angle the ring separation is prominent. Mars is more of a problem and to see it well a relatively high magnification must be used—up to 250 or 300 on a 15-centimetre reflector, for example.

It is important to remember that each time an image is enlarged, it becomes fainter and definition suffers. It is far better to study a smaller sharper picture than a larger diffuse one. Eyepieces are interchangeable from one telescope to another. The ideal is to have a whole series, ranging from the low-power wide-field type (for star clusters and similar views) up to high powers with small fields (for detailed observations of the Moon and planets). As soon as the image starts to become blurred a change should be made to a lower magnification eyepiece.

Especially when observing faint objects, "dark adaptation" is very important as the eye becomes much more sensitive to very faint impressions the longer it remains in the dark. Once exposed to bright light the adaptation will be lost until a further period has been spent in darkness. Very faint objects can often best be detected by directing the eye slightly to one side, while the observer's attention is concentrated on the spot where the object is believed to be.

Another problem concerns "extinction". When a star (or any other object) is low over the horizon, its light is received after passing through a relatively thick layer of the Earth's atmosphere. Since the atmosphere is dirty and turbulent, definition falls away and a star will twinkle strongly rather than remain a constant light. The effect of twinkling is thus due entirely to the atmosphere and has nothing to do with the stars themselves. If conditions are poor, it may well happen that a smaller telescope will give better results than a larger one.

Larger telescopes in amateur hands are very often of the reflecting type. A 30-centimetre reflector can reach stars below the 14th magnitude so that hundreds of variables come within range. Lunar craters less than 2 kilometres across can be detected, the disk of Jupiter is well delineated and most of Saturn's satellites and the ring system can be seen.

Any astronomer—amateur or professional—will inevitably tend to specialize and this will determine the choice of telescope. An observer of the Sun will certainly prefer a refractor to a reflector, while the planetary enthusiast will need as much magnification as possible. On the other hand a profitable, but laborious, field of research is to hunt for new comets and new exploding stars or novae. This means using a relatively low power but a wide field and some specialists use powerful mounted binoculars rather than telescopes.

The difference in resolution between three different telescopes can be seen in the drawings of Jupiter made by the author under identical viewing conditions. The view with a 7.5 cm refractor, was made using a magnification of 120 (×120); it shows the main belts as well as the Great Red Spot (*above left*). The view with a 15 cm reflector (×230) reveals even finer details (*above*). When using a 38 cm reflector the details of the belt system are clearly seen (*left*).

19

Observer's notebook

[April 18] 5in OG × 360.

			ω'	ω²	S.
27.	2347 est [Apr.12].	Centre of Red Spot.	...	009·5	-
28.	2357 est.	F. of Red Spot.	...	015·5	-
29.	0001 [Apr 18].	Projection from NEBs.	317·2	...	4
30.	0009.	Second part of prg. 29 from NEBs.	322·1	...	
31.	0011.	Projection from SEBs.	...	024·0	
32.	0015.	Small white spot in NTrZ.	...	026·4	
33.	0020.	F. of long projection 29/30 from NEBs.	328·8	...	
34.	0037.	C. of long dip in SEBs.	...	039·7	
35.	0047.	C. of slanting condensation in SEBs.	...	045·7	
36.	0052.	F. of condensation 35 in SEBs.	...	048·7	
37.	0103.	Small dip in NEBs. (Doubtful!)	355·0	...	
38.	0105.	F. of long dip 34 in SEBs.	...	056·6	
39.	0109.	Small white spot in NTrZ.	...	059·6	
40.	0111.	Small projection from NEBs.	359·9	...	
41.	0113.	F. of small white spot 39 in NTrZ.	...	061·4	5
42.	0147.	Small projection from NEBs. (Dubious!)	021·8	...	4 - 5
43.	0216 est.	White spot in STeZ.	...	099·5	
44.	0222 est.	White spot in EqZN.	043·1	...	
45.	0226 (12½ in. refl. × 200).	F. of white spot 43 in STeZ.	...	105·5	5 -

Apr.18, 0052. 5in OG × 330.
ω' = 348·3 ω² = 048·7 S = 4.

There was an extra belt in the STeZ, apparently in the latitude of the middle of the Red Spot.

The dark feature 35-6 was interesting. It formed part of the long, rather ill-defined dip in the S. border of the SEB, and was definitely angled at a slant. I could not see a white spot in the dip itself but this may have been due to the consistently bad conditions.

Star gazing will prove infinitely more satisfactory if an accurate and systematic record is made after each observation. A spiral notebook is best for this, or even a series of notebooks, one for each object. The notebook should always include the name of the observer, the type and aperture of the telescope, the magnification and the time of observation as well as the seeing conditions, based on the Antoniadi Scale. This scale is in roman numerals and ranges from I (perfect seeing) to V (very bad seeing). It is important to complete any drawings made at the telescope as soon as possible after observation. All temptation to "leave it until tomorrow" should be resisted, as errors of fact or interpretation are easily made and a faulty observation is worse than

RU PEGASI

k = 9.8

e = 9.0   h = 9.9   n = 11.2   r = 12.7   u = 13.7
f = 9.3   l = 10.5   p = 12.0   s = 13.1   w = 14.0
g = 9.5   m = 11.2   q = 12.5   t = 13.5   x = 14.2

1971

Date	Time		Instrument	Estimate	Mag	Comments
Jly 17	2240	1	12½in ×96	o1<m, o1<n	11.3	
Jly 20	0020	1	"	≈p, 0.5>q	12.0	
Jly 21	2330	5	" ×116	=q	12.5	Clouding.
Jly 26	0040	1	"	o1<q	12.6	
Jly 29	2300	1	"	=q	12.5	
Aug.15	2340	4	"	=q	12.5	
Aug.16	2350	1	"	o1<q	12.6	
Aug.24	2300	1	"	=q	12.5	
Aug.26	2300	1	"	o.1>q	12.4	Also with 5in OG.
Aug.31	2010	1	" ×96	o.1>l, o.7>m.n	10.4	
Set.19	2200	4	" ×116	≈q	12.5	
Set.20	2020	4	"	=q	12.5	
Set.21	1940	2	"	=q	12.5	
Set.22	1930	5	" ×96	=q	12.5	Misty.
Set.25	2220	3	" ×116	≈q	12.5	
Set.26	2200	1	5in OG ×116	=q	12.5	
Set.28	2140	1	12½in ×116	=q	12.5	
Oct.5	2625	5	"	≈q	≈12.5	Moon.
Oct.6	2010	4	"	=q	12.5	"
Oct.7	1900	3	"	=q	12.5	"
Oct.10	1910	1	"	≈q	12.5	
Oct.12	2110	3	"	=q	12.5	Mist.
Oct.17	2030	3	"	=q	12.5	
Oct.19	2100	1	"	=q	12.5	
Oct.20	2100	3	"	=q	12.5	
Oct.22	2200	2	"	o1<q	12.6	
Oct.24	1940	1	"	=q	12.5	
Oct.26	2000	1	"	>l, o.7>m	10.5	

useless. The timing of observations should always be in Greenwich Mean Time (GMT), which is known as Universal Time (UT) by astronomers.

The extract from the author's Jupiter notebook (*above left*) shows a sketch of that planet and six columns filled with notation; there is also a comments column.

The second extract (*above right*) is from the author's notebook on variable stars. RU Pegasi is a dwarf nova or SS Cygni star that is normally of magnitude 12.5 but, as has been noted, every now and again it brightens to magnitude 10. A list of ordinary stars has also been included (column 5) so a comparison can readily be made with the estimated magnitudes of RU Pegasi (column 6).

The Moon 1

The Moon's origin is still a matter for debate but few astronomers now believe that it broke away from the Earth, as used to be thought. They hold rather that the Moon has always been an independent body.

The Moon is officially ranked as the Earth's satellite but, since it is relatively large and massive, with a diameter of 3,476 kilometres and a mass of 0.012 that of the Earth, it may better be regarded as a companion planet. The mean distance from the Earth is 384,400 kilometres and the Moon's low escape velocity (2.4 kilometres per second) means that it has been unable to retain any appreciable atmosphere. Indeed, analysis of the lunar material brought back by the Apollo astronauts and the Russian automatic probes has confirmed that no life has ever existed there.

The Moon shines only by reflected sunlight although, when in the crescent stage, the "dark" hemisphere may often be seen shining faintly because of light reflected from the Earth. Splendid though the Moon may appear in the sky, it is in fact a very junior member of the Solar System.

The Moon takes 27.3 days to complete one revolution in its orbit. It takes exactly the same time to spin once on its axis

The Moon has one-quarter the diameter and 0.012 the mass of the Earth. It has a small core, which is considerably less, in proportion, to that of the Earth's core. Because its centre of gravity is eccentric, its crust is thinner on one side than the other and it is also thinner under the numerous maria.

The familiar lunar phases occur because the Moon does not always turn its daylight side towards the Earth. With the Sun's rays coming from the right, the Moon is new (1) and its dark side is turned towards the Earth. It cannot then be seen unless the alignment is perfect enough to produce a solar eclipse. At 2 it is half (first quarter); at 3 full; and at 4 half once more (last quarter). Between half and full phases the Moon is "gibbous".

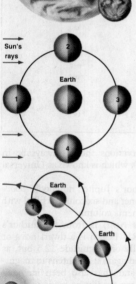

The Moon takes 27.3 days to move round the barycentre—the centre of gravity of the Earth–Moon system, which lies within the Earth's globe. However, the Earth is moving round the Sun. The Moon (*left*) is new at 1; when it has returned to 1, during the Earth's orbit of the Sun, it is still not lined up and must move on to 2 before it is new again. The "lunation", or interval between successive new moons, is therefore 29.5 days not 27.3.

(this is known as captured or synchronous rotation). The rotation of the Moon has been slowed down by the action of the Earth until by now it keeps the same face towards Earth. However, the Moon does not keep the same face towards the Sun, so that day and night conditions are the same in each hemisphere. To an observer on the Moon's "near" side, the Earth seems to stay almost motionless in the lunar sky; from the "far" side the Earth would never be seen.

The Moon's orbit is not circular; it moves fastest when closest to the Earth so that the position in orbit and the amount of axial spin become "out of step". The Moon thus seems to sway slightly, allowing observers to see beyond alternate limbs (libration in longitude). This means that from Earth 59 per cent of the total lunar surface may be seen but never more than 50 per cent at any one moment. The remaining 41 per cent remained unknown until the circum-lunar flight of the Lunik 3 in 1959. Since then there have been the Apollo missions, resulting in the first manned landing of Apollo 11 in 1969. In particular, Apollo 15, 16 and 17 included detailed geological surveys providing knowledge of the formation of the crust, mantle and core.

Lunar eclipse	Max. phase	Type and duration
28 Oct. 1985	17 hrs. 44 m.	Total, for 42 minutes
24 Apr. 1986	12 hrs. 44 m.	Total, for 68 minutes
17 Oct. 1986	19 hrs. 19 m.	Total, for 74 minutes
7 Oct. 1987	03 hrs. 59 m.	Partial, 1% eclipsed
27 Aug. 1988	11 hrs. 06 m.	Partial, 30% eclipsed

Three photographs of a lunar eclipse, taken before totality, show the edge of the Earth's shadow crossing the lunar surface. During an eclipse, the Moon does not disappear completely because light is refracted on to its surface by way of the Earth's atmosphere. The period of totality can be as much as 1¾ hours.

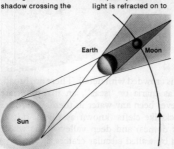

An eclipse of the Moon occurs when the Moon passes into the cone of shadow cast by the Earth. If the Moon partially enters the cone there is a partial eclipse; if it wholly enters the cone the eclipse is total. Eclipses do not happen at every full Moon because the lunar orbit is appreciably inclined.

The Moon 2

Page 32
Page 28

Page 26
Page 30

The Moon's surface is crowded with detail. There are broad, grey plains known as maria or "seas", an inappropriate name as there has never been any water in them. There are also mountains, crack-like clefts (known as rilles), ridges, swellings (known as domes) and deep valleys. The whole surface is dominated by walled circular craters.

24

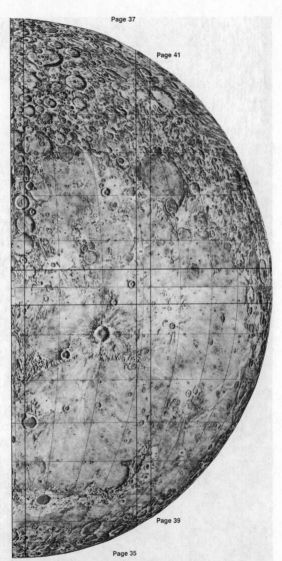

Page 37

Page 41

Page 39

Page 35

The craters and walled plains range from enclosures over 200 kilometres across to tiny pits at the limit of visibility. Some are perfectly circular; others have been broken and distorted. Their origin is still a matter for debate. Some astronomers regard their origin to be volcanic, while others believe they have been caused by the impact of meteorites.

25

Far north-east sector

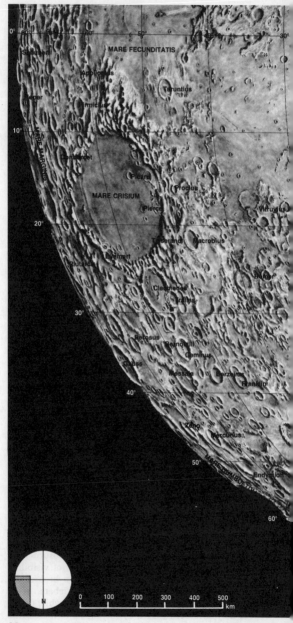

MARE FECUNDITATIS

Apollonius

Firmicus

Taruntius

Condorcet

Picard

Proclus

MARE CRISIUM

Pierce

Vitruvius

Azophi

Tisserand

Macrobius

Eimmart

Hahn

Plutarch

Cleomedes

Tralles

Berosus

Hahn

Geminus

Gauss

Messala

Berzelius

Franklin

Zeno

Mercurius

MARE HUMBOLDTIANUM

Endymion

N

0   100   200   300   400   500
km

26

Mare Crisium, near the terminator, is shown when the Moon is just past full. Cleomedes, with its darkish floor, and Tralles stand out clearly to the south.

A more favourable libration means that the Mare Crisium is further on to the disk and appears less foreshortened, making the surface details more clear.

Proclus, the ray centre to the west of the Mare, contains a low central mountain. The crater is not obvious under low illumination. It is 29 km in diameter and 2,500 m deep.

Under direct illumination the ray system of Proclus is very clear. The rays do not cross the darkish area of Palus Somnil, which leads on to the Mare Tranquillitatis.

This region of the Moon is dominated by the Mare Crisium, which is one of the smaller lunar "seas". It is one of the most distinctive and is easily visible with the naked eye. Its appearance is deceptive as it actually measures 560 kilometres in an east-west direction but only 448 kilometres north-south. The surface is comparatively smooth, but there are two well-marked small craters, Picard and Pierce. Outside the Mare lies Proclus, which is one of the brightest craters on the Moon and is the centre of an asymmetrical system of bright rays. Cleomedes has walls with peaks rising to over 2,793 metres.

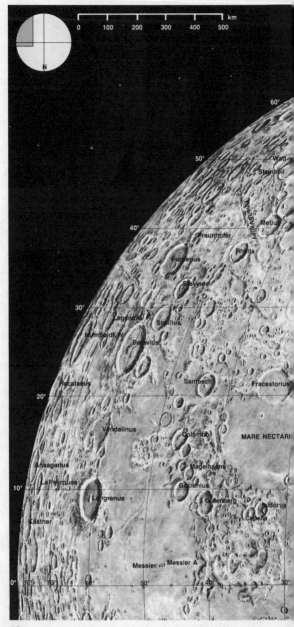

km
0 100 200 300 400 500

N

60°

50° Watt
Steinheil

Rheita Valley

40° Metius
Fraunhofer
Furnerius Rheita

Stevinus

30° Legendre Snellius
Humboldt N
Petavius

Hecataeus Santbech Fracastorius

20°
Vendelinus
Colombo
MARE NECTARI

Madelhaans
Ansgarius
La Peyrouse Goclenius
10° Langrenus Gutenberg Isidorus
Kästner Capella

Messier Messier A
0° 80° 70° 60° 50° 40° 30°

Petavius is one of the finest craters on the Moon. It has a slightly convex floor and a grand central mountain group from which a magnificent rille runs to the south-west wall.

Messier and Messier A are two small craters from which extends a double ray that makes the twins look curiously like a comet. Messier A is the larger and less regular of the two.

Fracastorius is a bay situated on the southernmost point of the Mare Nectaris. The seaward wall has been almost destroyed and there is a dark streak across the floor.

The Rheita Valley is 185 km from one end to the other and 24 km across its widest part. It is in fact not a true valley but a string of minute craters and unremarkable rings.

This region of the Moon contains some majestic walled plains. Langrenus is a noble formation with high, terraced walls and a central peak; Petavius is of the same type and Vendelinus is similarly sized but is less well preserved. Close to the limb are more large enclosures, such as Wilhelm Humboldt. Furnerius is another large enclosure, which is fairly well preserved. The region includes parts of the Mare Fecunditatis and the Mare Nectaris, including Fracastorius. The Messier twins have strange "comet" rays and are easy to identify and there is also the so-called Rheita Valley, which is really a crater chain.

29

MARE TRANQUILLITATIS

Sabine
Ritter
Ariadaeus
Julius Caesar
Boscovich
Sosigenes
Agrippa
Godin
Rhaeticus
Triesnecker
Hyginus

Plinius
Menelaus
Manilius
MARE VAPORUM

Bessel

Littrow
Le Monnier
Chacornae
Posidonius
Lacus Somniorum
MARE SERENITATIS
Linné
Palus Putredinus
Autolycus
Aristillus
Caucasus Montes
Calippus
Cassini

Williams
Hercules
Atlas
Lacus Mortis
Burg
Eudoxus
Aristoteles
Alpine Valley

Mitchell
Bailly
Ingles
Drabo
Bond, W
Meton

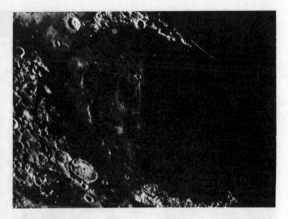

The Mare Serenitatis has a relatively smooth floor but contains several craterlets. There are many prominent ridges, some of which seem to be the walls of ghost craters. A controversial bright ray runs across the Mare.

Linné achieved notoriety in the 1860s because of its alleged change in form; this change is now discounted. Linné is in fact a small craterlet surrounded by a light patch.

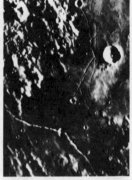

Triesnecker is associated with an extensive rille system and the Hyginus "rille" (really a crater chain) is a quick and easy object to pick up under a small telescope.

This section includes part of the Mare Imbrium, including Aristillus and Autolycus, which belong to the Archimedes group. One of the most regular of all the lunar "seas" is the Mare Serenitatis, which contains no large formations, although there is one small, bright crater, Bessel, and a small craterlet, Linné. North of the Mare Serenitatis are the great walled formations Aristoteles and Eudoxus and the deep, smaller crater Bürg. To the south, Triesnecker and Hyginus, which are part of the Mare Vaporum, are visible. Boscovich and Julius Caesar are distinguishable by their dark floors and Plinius is a superb crater.

31

km
0   100   200   300   400   500

N

Maginus

Pontécoulant

Vlacq

Lilius

Pitiscus

Baco

Cuvier

Janssen

Barocius

Faraday

Maurolycus

Stöfler

Miller

Rabbi Levi

Gemma Frisius

Lindenau

Walter

Aliacensis

Piccolomini

Werner

Fermat

Apianus

Krusenstern

Blanchini

Polybius

Sacrobosco

Playfair

Azophi

Abenezra

Abenezra

Catharina

Almanon

MARE
NECTARIS

Tacitus

Cyrillus

Abulfeda

Theophilus

Descartes

Albategnius

Zöllner

Halley

Hipparchus

Torricelli

Saunder

Horrocks

Delambre

30°   20°   10°

Albategnius is a vast crater with quite high walls and a few peaks rising to over 3,000 m. It forms a notable pair with Hipparchus, a large enclosure containing a wealth of fine detail.

One of the grandest craters on the Moon is Theophilus with its walls rising to over 5,000 m. It has a great multi-peaked central mountain mass and terraced inner ramparts.

The Altai Scarp is part of the ring system of the Mare Nectaris. The Scarp rises steeply in the east but is only slightly above mean ground level in the west.

Stöfler, an enclosure that measures 145 km across, has a darkish floor, which makes it easy to identify. Part of the rampart has been destroyed by the intrusion of Faraday.

This section contains few "seas" and most of the area is occupied by rugged uplands. A variety of craters and walled plains may be seen, including the huge ruined Janssen, with a diameter of over 160 kilometres, and superb formations such as Theophilus, which is the northern member of a chain of three great walled plains and actually intrudes into its neighbour Cyrillus. There are numerous small rings and countless craterlets. Lofty mountain ranges are absent but there is an interesting feature known as the Altai Scarp running north-west from Piccolomini. The Scarp rises to about 1,800 metres above mean ground level.

North-west sector

Archimedes is very regular and is 80 km in diameter with fairly low walls. The slightly sunken floor is darkish and very smooth with no vestige of a central mountain.

Copernicus, which is nicknamed the "Monarch of the Moon", is 90 km across and has massive, terraced walls. Its rays dominate this whole part of the lunar surface.

Eratosthenes ranks as one of the most perfect craters on the Moon. It is 61 km across, has much floor detail and forms a triangle with Copernicus and ruined Stadius.

Sinus Iridum, the lovely "Bay of Rainbows", is one of the most striking sights on the Moon near sunrise, when the floor is in shadow and the Jura Mountains are illuminated.

This is one of the most interesting regions on the Moon. It contains most of the Mare Imbrium as well as the superbly beautiful Sinus Iridum. The Apennines, ending near the deep, prominent crater Eratosthenes, make up part of the border of the Mare; the Carpathians are rather less lofty. On the Mare Imbrium itself are many small formations as well as some major enclosures of which Archimedes is pre-eminent. The regular formation Plato is in the region of the Alps. To the south of the area lies Copernicus and near it is the famous ghost crater Stadius. Other interesting features are Beer and Feuillée and the mountains Pico and Piton.

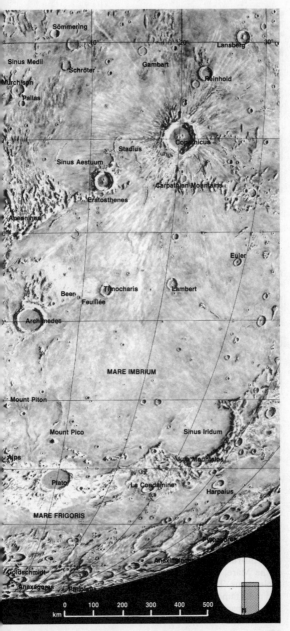

Sömmering

Sinus Medii

Murchison

Pallas

Schröter

Gambart

Lansberg

Reinhold

Copernicus

Stadius

Sinus Aestuum

Eratosthenes

Carpathian Mountains

Apennines

Euler

Beer

Timocharis

Lambert

Feuillée

Archimedes

MARE IMBRIUM

Mount Piton

Mount Pico

Sinus Iridum

Alps

Jura Mountains

Plato

Le Condamine

Harpalus

MARE FRIGORIS

Anaximenes

Goldschmidt

Anaxagoras

Barrow

| 0 | 100 | 200 | 300 | 400 | 500 |
km

N

35

South-west sector

Clavius, one of the largest of the so-called walled plains, has a chain of craters, arranged in a curve, crossing its floor. It is 233 km in diameter and has walls rising to over 3,600 m.

At the centre of the greatest ray system on the Moon is Tycho, a well-formed crater, 87 km across, with high terraced walls and a central mountain complex.

Ptolemaeus, Alphonsus and Arzachel form a great chain of walled plains known as the Ptolemaeus chain. The gradation in type between these three is very interesting and significant.

The Straight Wall is not straight, nor is it a wall; it is a particularly interesting fault, 244 m from crest to foot, that lies between Thebit and Birt and ends in a group of hills.

Some of the most famous features of the Moon are to be found in this sector. Tycho in the uplands is well formed and there are many other large structures such as Clavius, Maginus, Longomontanus and the compound Schiller. To the north lies much of the Mare Nubium, on which there is the semi-ruined Fra Mauro group. Bullialdus, near the edge of the Mare, is a well-formed crater and at the far side is Ptolemaeus. Thebit is a perfect example of a multiple crater and most famous of all lunar faults is the Straight Wall, near Birt. Purbach, on the edge of the region, is associated with two more major formations, Regiomontanus and Walter.

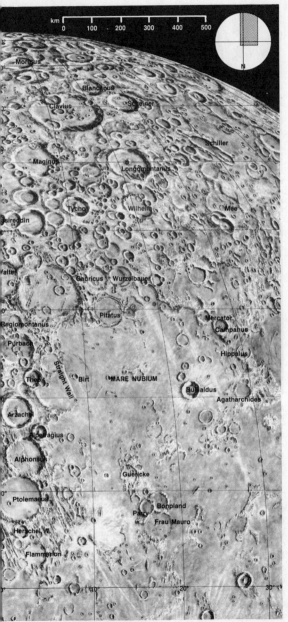

km
0   100   200   300   400   500

N

Morchus
Blancanus
Clavius
Scheiner
Schiller
Maginus
Longomontanus
Tycho
Wilhelm
Mee
asireddin
Valter
Gauricus   Wurzelbauer
Pitatus
Regiomontanus
Mercator
Campanus
Purbach
Hippalus
Thebit
Birt
MARE NUBIUM
Bullialdus
Straight Wall
Agatharchides
Arzachel
Alpetragius
Alphonsus
Guericke
Ptolemaeus
Bonpland
Parry
Herschel
Frau Mauro
Flammarion

0°   10°   20°   30°

37

Far north-west sector

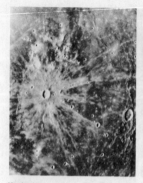

Kepler has a central mountain, interior radial bands and walls that are so heavily terraced that they seem to be double in places.

Hevelius, which is a member of the great Grimaldi chain, has a convex floor containing a central elevation and a system of rilles.

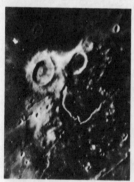

Aristarchus is the most "active" and brightest formation on the Moon. Together with Herodotus, Prinz and Schröter's Valley, it has produced more than half the TLP (Transient Lunar Phenomena) reported. Herodotus has a dark floor, in striking contrast to its neighbour.

Otto Struve is a vast enclosure made up of two old rings, each about 160 km across, which have now merged. The west wall is really a ridge parallel with the mountains beyond. Near by and easy to locate is Briggs, which is connected to Seleucus by a series of ridges.

This area of the Moon is dominated by the Oceanus Procellarum, much the largest although by no means the best-defined of the lunar "seas". The most interesting group is undoubtedly that of Aristarchus. Adjoining it is Herodotus, of similar size but much less brilliant, and extending from Herodotus is Schröter's Valley. This whole region is subject to the elusive glows known as Transient Lunar Phenomena or TLP. The Harbinger Mountains, near Aristarchus, are more like clumps of hills than a major range. Along the limb are some large, generally rather low-walled formations and the area also includes Hevelius.

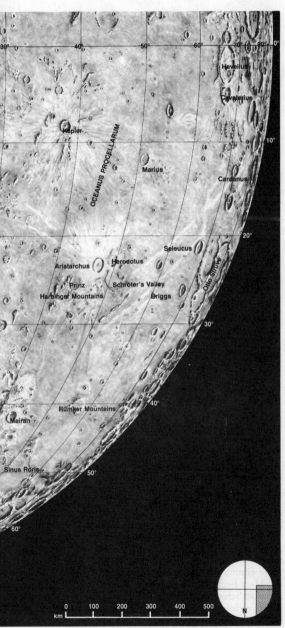

30° 40° 50° 60° 70° 80° 0°

Hevelius

Cavalerius

Kepler

0°

OCEANUS PROCELLARUM

Marius

Cardanus

20°

Seleucus

Herodotus

Aristarchus

Prinz

Schröter's Valley

Harbinger Mountains

Briggs

Otto Struve

30°

Rümker Mountains

40°

Mairan

Sinus Roris

50°

60°

0 100 200 300 400 500
km

N

Far south-west sector

The Mare Humorum is a superb minor circular "sea" and it provides the most striking example of faulting on the Moon. There are also many rilles and bays in and near it.

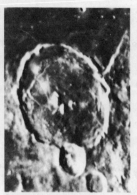

Gassendi, which is 88 km in diameter, is an important walled formation. The floor includes a central peak and a magnificent system of rilles that is easily seen from Earth.

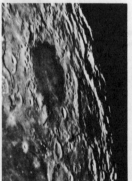

Grimaldi is one of the grandest enclosures on the Moon, so too is its neighbour, Riccioli. Both have dark patches on their floors and irregular walls. Many TLP have been reported.

Wargentin represents the only example of a really large well-preserved lunar plateau. It is 88 km in diameter and has various hills and ridges on its plateau.

The most famous walled formation in this region is probably Grimaldi; adjoining it is Riccioli. Not far from Grimaldi is Sirsalis, a double crater associated with one of the finest rilles on the Moon. The only "sea" area in the region is the Mare Humorum; at its north border is Gassendi, which has a low "seaward" wall. Doppelmayer, to the south of the Mare Humorum, is in even worse repair, the "wall" facing the Mare being barely traceable. To the far south lies the majestic Schickard and the celebrated plateau Wargentin, which is close to the limb and very foreshortened, but it is not hard to identify with a small telescope.

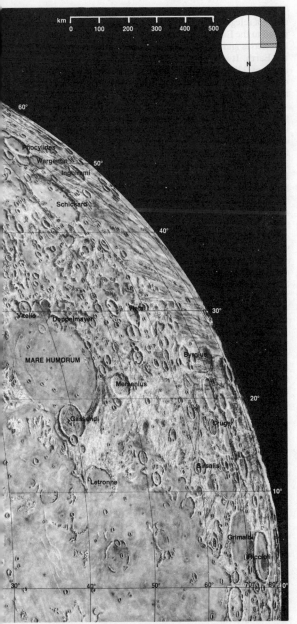

km
0   100   200   300   400   500

N

60°

Phocylides
Wargentin
Inghirami   50°

Schickard

40°

Vitello   Mee   30°
Doppelmayer

MARE HUMORUM   Byrgius

Mersenius   20°

Gassendi   Crüger

Sirsalis

Letronne   10°

Grimaldi

Riccioli

30°   40°   50°   60°   70° 80° 0°

41

The Sun 1

The Sun is a star of no particular importance; indeed, astronomers relegate it to the status of a yellow dwarf. Yet, compared to the Earth, it is a vast globe. Its diameter is 1,392,000 kilometres and its volume is well over a million times that of the Earth. It is a typical member of the Galaxy and takes as much as 225 million years to make one complete revolution of the galactic centre, along with the Earth and all other members of the Solar System.

The surface temperature of the Sun is 6,000 degrees centigrade. Near the core, the temperature rises to at least 14 million degrees centigrade. The Sun is not "burning" in the conventional sense. Deep inside its globe, nuclear transformations are taking place. Hydrogen is being converted into helium and each time a new helium nucleus is formed from four hydrogen nuclei a little energy is released and a little mass is lost. This energy keeps the Sun shining, while the mass-loss amounts to four million tonnes per second. Although the Sun is halfway through its lifetime, it still has sufficient fuel for another five thousand million years.

The Sun is a fascinating, although dangerous, object to study. *An observer must never look directly at the Sun through a telescope or even binoculars.* The inevitable result will be permanent blindness. Nor is it safe to look directly at the Sun merely using a dark filter over the eyepiece. The only sensible way to view the Sun's surface is to project it through a telescope. Refractors are ideal for all solar work, although reflectors may also be used.

The most obvious features on the solar surface are the sunspots, which appear as dark patches. A typical large sunspot has an umbra, or central area, surrounded by a lighter area known as the penumbra. No sunspot can last for more than a few months and many have lifetimes of only a few days; when the Sun is at its most active many groups can be seen at the same time.

As the Moon passes between the Sun and the Earth, there is a solar eclipse. If the eclipse is total, the Sun's outer surroundings flash into view and the appearance is magnificent. Unfortunately solar eclipses do not occur every month because the Moon's orbit is inclined to that of the Earth and, at most new Moons, the Moon passes either "above" or "below" the Sun in the sky, thereby avoiding eclipse.

To project the Sun's image safely, point the telescope sunwards. Without looking through the eyepiece, remove the cap from the object glass; the Sun's image may be thrown either on to a screen inside a simply made black box (*right*) or on to a sheet of white paper held behind the eyepiece.

Eclipses of the Sun occur because the Moon appears almost the same size as the Sun in the sky. If the Earth, Moon and Sun are aligned exactly, the Moon will blot out the Sun, causing a total eclipse (1). Because the lunar shadow only just reaches the Earth, the zone of totality is narrow. More often the Moon never fully eclipses the Sun and a partial eclipse occurs (2). In an annular eclipse (3) the Moon is near the far point of its orbit and cannot cover the Sun completely.

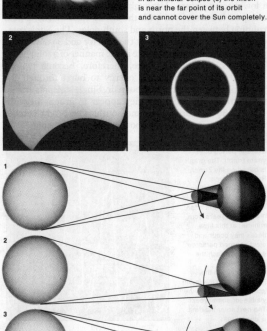

Eclipse	Type	Duration	Extent of visibility
12 Nov. 1985	total	1 m. 58 s.	S. Pacific, Antarctic
9 Apr. 1986	partial		Antarctic
3 Oct. 1986	total	0 m. 3 s.	N. Atlantic
29 Mar. 1987	total	0 m. 56 s.	Argentina, Atlantic, Central Africa, Indian Ocean
23 Sept.1987	annular		
18 Mar. 1988	total	3 m. 46 s.	Indian Ocean, E. Indies, Pacific
7 Mar. 1989	partial		Arctic
31 Aug. 1989	partial		Antarctic

The Sun 2

The visible surface of the Sun is known as the photosphere. This emits virtually all the Sun's radiation and has a temperature of 6,000 degrees centigrade. Through a small telescope the photosphere will have a mottled appearance, larger telescopes refine this to reveal granulation—the effect of convection in the outer layers of the Sun.

An interesting and easily observable phenomena to study on the Sun are dark patches known as sunspots, which only appear dark in comparison to the photosphere as they have a lower temperature. Although not completely understood, sunspots are associated with the strong solar magnetic field running between the north and south poles. The Sun's rotation varies between 26 days at the equator and 36 days at the poles because it does not rotate in the manner of a solid body. The lines of magnetic force, therefore, become twisted causing a loop of magnetic energy to burst through the photosphere and form a sunspot pair. Single sunspots are not uncommon, but generally they appear in groups between 10° and 30° latitude north and south. The bright patches usually associated with sunspots are known as faculae.

The solar cycle has an average period of 11 years (*right*). The graph is based on the Zurich Relative Sunspot Number, for recording sunspots. Maximum sunspot activity is often reached 4.5 years from minima; at this time flares may occur and eject charged particles which may reach the Earth causing disturbances in the ionosphere. At minimum activity no spots may be visible for some weeks. The next solar maxima is due in 1991.

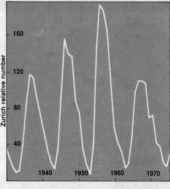

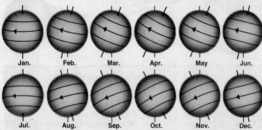

| Jan. | Feb. | Mar. | Apr. | May | Jun. |

| Jul. | Aug. | Sep. | Oct. | Nov. | Dec. |

Sunspots travel east to west across the solar disk but their paths do not always follow a straight line and more often appear curved. This effect is due to the inclination to the plane of the ecliptic of the Earth's rotational axis, which is $23\frac{1}{2}°$. Thus different aspects of the Sun are presented to the Earth during its revolution around the Sun. The paths of sunspots only appear as a straight line in December and June.

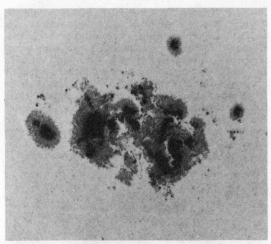

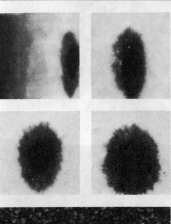

Sunspots are relatively cool patches that appear as dark spots on the photosphere (*above*). The dark central region, or umbra, is surrounded by a fainter area known as the penumbra. Sunspots are not completely understood but are thought to be associated with the Sun's strong magnetic field. The Wilson Effect (*left*) occurs when a sunspot reaches the limb. The penumbra then appears narrower on one side, indicating that the spot is saucer-shaped.

The Sun's visible surface or photosphere (*left*) has a granular appearance when viewed through a moderate-sized telescope under good conditions. This effect is caused by convection: columns of hot gas circulate upwards in the centre of the granule and descend as cool gas at the boundary. Each granule is the apex of a column ranging from approximately 1,200 km to 30,000 km in diameter.

45

The Sun 3

Simple telescopic observations of the Sun show only the bright surface, or photosphere, and features such as spots and faculae. The photosphere is surrounded by the chromosphere, in which violent phenomena known as flares occur; these emit charged particles and short-wave radiation. Beyond this comes the inner κ corona, made of extremely tenuous gas. Both the chromosphere and κ corona can be seen with the naked eye only during a total eclipse; they are best studied with equipment based on the principle of the spectroscope, involving observation in the light of one element only—usually hydrogen or calcium. The outer F corona extending far into space is hard to observe; it is distinguished by the dark absorption lines in its spectrum.

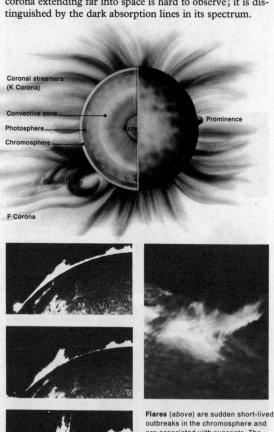

Coronal streamers
(K Corona)

Convective zone

Photosphere

Chromosphere

Prominence

Coronal Cap

F Corona

Flares (*above*) are sudden short-lived outbreaks in the chromosphere and are associated with sunspots. The sequence (*left*) shows a flare surge and filament disruption over a 30-minute period. The surge of charged particles emitted may give rise to aurorae and magnetic storms.

Prominences are outbursts of incandescent gas rising from the chromosphere. They are of two main types, quiescent and eruptive. Quiescent prominences, such as the hedgerow type (*above*), are relatively stable and may persist for days or months before dispersing. Eruptive prominences are more violent. The example (*left*) extends to about 920,000 km from the Sun but many reach to at least two million kilometres. It is possible to observe prominences with a spectroheliograph, an instrument for photographing the Sun in the light of one element only. When the Sun is examined in hydrogen light, prominences are seen against the bright disk of the Sun as dark filaments, but when photographed in the light of calcium the flocculi (or plages) will appear bright.

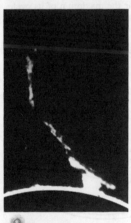

Present knowledge of the Sun is obtained using instruments based on the principle of the spectroscope—an instrument for studying the spectrum of light. An incandescent solid, liquid or gas at high pressure produces a continuous spectrum (*top*), from red at the long-wavelength end through to violet at the short-wavelength end. An incandescent gas at low pressure produces an emission spectrum of isolated bright lines (*centre*); each is characteristic of an element or group of elements. Thus if two bright yellow lines are seen in a particular position they must be caused by sodium. The solar spectrum (*bottom*) shows the combined effect of the photosphere yielding a continuous background spectrum and the chromosphere or "reversing layer" of the Sun producing dark absorption lines which if seen by themselves would appear bright in emission. The dark lines are named after Josef von Fraunhofer (1787–1826).

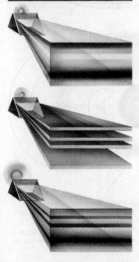

The Sun 4

In addition to sending the Earth virtually all its light and heat, the Sun has effects that are much less obvious. Of special importance is the so-called solar wind, made up of charged particles streaming continuously outwards from the Sun in all directions. As it passes the Earth it is moving at about 600 kilometres per second. The solar wind affects the magnetosphere of the Earth (the area inside which the Earth's magnetic field is dominant). The particles also enter the radiation zones around the Earth known as the Van Allen Belts. At times of great activity there is interference with some wavelengths of radio reception because of disturbances in the ionosphere (the region of the Earth's upper air in which radio waves are reflected). These effects are most noticeable when the Sun is near the maximum of its cycle. Apart from visible light and heat, the Sun is also a source of radio waves, ultra-violet radiation and x-rays.

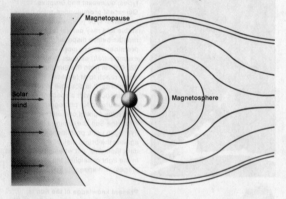

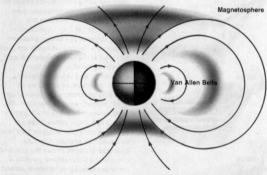

The magnetopause (*top*), which is the outermost boundary of the magnetosphere, is the shock front that meets the charged particles emitted from the Sun. Within this region lie the Van Allen Belts (*above*); these zones, which look rather like doughnuts, capture the charged particles. It is the leakage of particles into the upper atmosphere that causes spectacular auroral displays.

The Sun

▲2

▼3

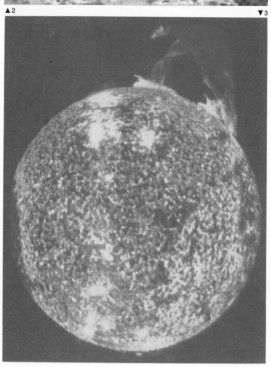

1 The outer corona predominates in this total eclipse
2 A solar eruption taken from Skylab using extreme ultraviolet.
3 A solar prominence, 400,000 km high, as seen from Skylab.
4 The inner corona and prominences predominate in this total eclipse.
5 An eruptive prominence seen from Skylab.

▼5   ▲4

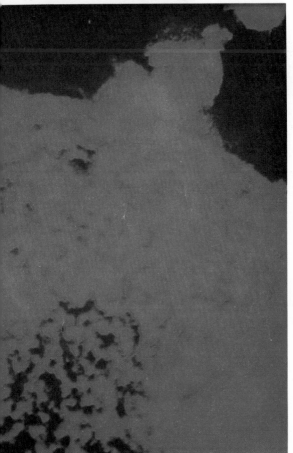

The Earth and Moon

6 The rising Earth
viewed from Apollo 8.
7 This view, from
Apollo 8, illustrates the
similarity between the
near and far sides of
the Moon. Mare Crisium
is the dark, oval sea
dominating the area
near the terminator.
Tsiolkovskii, the far-
side crater with a dark
floor and prominent
central peak, is near the
limb, to the upper left.
8 During an eclipse
of the Moon, the lunar
surface looks coppery.

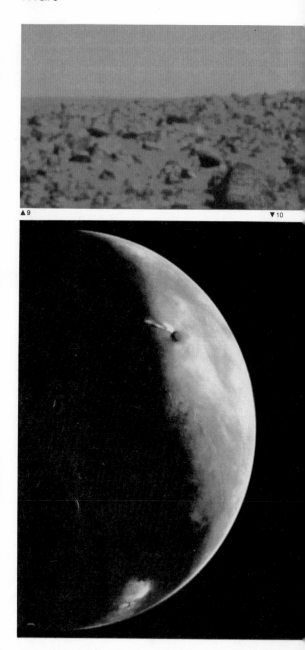

▼ 11

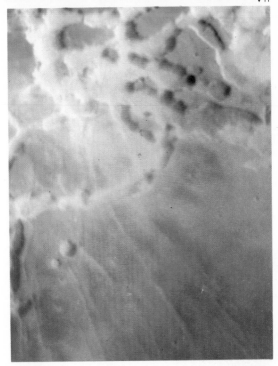

9 The Martian surface, taken from Viking 1, shows orange-red materials covering most of the surface and darker bed rock exposed in patches. **10** Viking 2 approached Mars from the dark side, giving this crescent view. The prominent volcano at the upper left is Ascreaus Mons. **11** The bright clouds of water ice are in and around Noctis Labyrinthus, the tributary canyons in a high plateau region on Mars.

Jupiter and Saturn

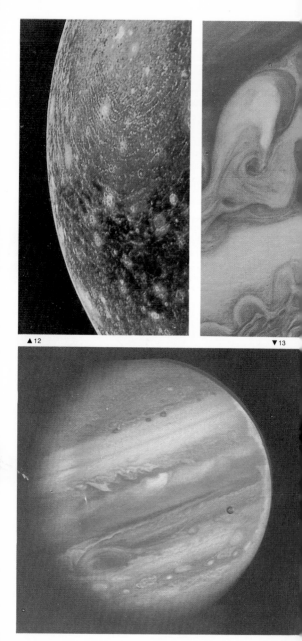

▲ 12

▼ 13

▲ 14

▼ 15

12 Callisto, one of the Galilean satellites, apparently consists of a mixture of ice and rock.
13 Io (left) and Europa stand out dramatically against the disk of Jupiter, as seen from Voyager 1.

14 The region of the Great Red Spot is very violent. A white oval, visible from Earth, is prominent.
15 This view of Saturn, from Catalina Observatory, shows the contrasting brightness of the rings.

Clusters and Nebulae

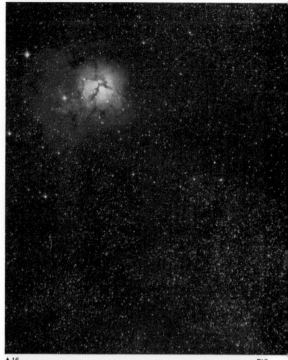

▲ 16   ▼17

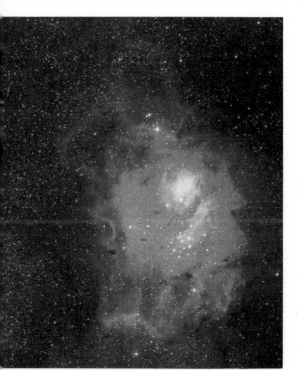

▼18

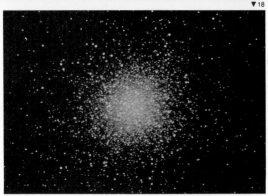

16 The Sagittarius region is particularly rich in nebulae and clusters. Here two typical emission nebulae—the Trifid Nebula, M20 (left) and the Lagoon Nebula, M8— are shown. Both are radio sources.

17 One of the most spectacular clusters is the Pleiades, M45, in Taurus. This detail emphasizes the associated nebulosity.
18 The finest globular in the northern sky is M13 in Hercules.

Nebulae

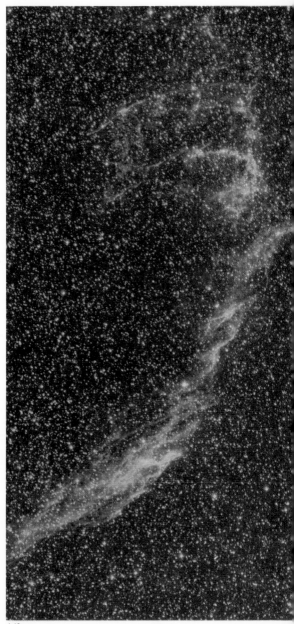

▲ 20

▼ 21

19 The Veil Nebula, NGC 6992, in Cygnus, is probably the result of a supernova explosion that occurred in prehistoric times. It is estimated that it will cease to be luminous in 25,000 years' time.

20 Unlike any other nova, Eta Carinae is intensely luminous and associated with complex nebulosity. **21** The Rosette Nebula, NGC 2237, is a typical emission nebula in the faint constellation Monoceros.

Galaxies

▲ 23

22 The Great Spiral in Andromeda, M31, is a member of the Local Group. It is 2.2 million light-years away.
23 In this view of the nucleus of the Great Spiral, taken at the correct exposure, the arms are not well seen.
24 M82, in Ursa Major, is the best-known example of an exploding galaxy; it is also a strong radio source. It lies at a distance of 10 million light-years.
25 The Whirlpool Galaxy, M51, in Canes Venatici, is 37 million light-years away. It was the first galaxy to be seen as a spiral (by Lord Rosse in 1845).

▲ 24   ▼ 25

Aurorae

▲ 26

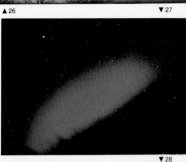

▼ 27

26 This brilliant
display of aurorae is
typical of those seen
at high latitudes.
27 Aurorae can take
many different forms.
This type is often
called a ''flaming surge''
28 This dramatic view
was photographed from
a NASA aircraft
while being used as
an observation platform
to study the effects of
the Aurora Borealis.

▼ 28

Aurorae, or Polar Lights, are a spectacular and beautiful phenomena (*left*) which occur as a result of activity on the Sun. Aurorae observed in the Northern and Southern Hemispheres are called Aurora Borealis and Aurora Australis respectively. Although this phenomena is not completely understood it is thought that charged particles emitted from the Sun (solar wind) are trapped in zones around the Earth, known as Van Allen Belts. These zones become overloaded and the particles, accelerated by the Earth's magnetic field, cascade downwards giving rise to aurorae. They may take various forms, the commonest being the "auroral curtain". Other displays may be arcs, rays, streamers and flaming surges. On clear nights it is possible to observe some form of display between the latitudes of 60° and 75° in each hemisphere. They usually occur at heights ranging from 100 to 1,000 kilometres.

Zodiacal Light (*below*) is one of various "sky glows", which should not be confused with aurorae. Zodiacal Light is seen as a faint cone of light rising from the horizon along the ecliptic before sunrise or after sunset. It is best seen when the ecliptic is most nearly perpendicular to the horizon during the months of February, March, September and October. It originates outside the Earth's atmosphere and is due to light being reflected from dust grains in the main plane of the ecliptic. The particles measure 0.1 to 0.2 microns (one micron is one-millionth of a millimetre).

A much more elusive phenomenon is the Gegenschein or Counterglow, a faint patch of light exactly opposite to the Sun in the sky. Its diameter may be up to 40 times that of the full Moon, but it is never easy to see.

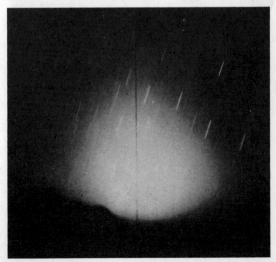

Mercury

Mercury is the smallest of the principal planets (with the possible exception of Pluto) and, although it can become brighter than any star, it is not very easy to see with the naked eye. It is never visible against a dark background and always keeps close to the Sun in the sky so that, without optical aid, it is only seen low in the west after sunset or low in the east before sunrise. Telescopically, all that can be made out is the characteristic phase. Before the flight of Mariner 10 in 1974 little was known of its surface.

Mariner 10, which made three active passes of Mercury between 1974 and 1975, showed that the surface is heavily cratered, looking superficially very much like that of the Moon. There are mountains, valleys, scarps and ridges. There are also depressed basins, of which one (the Caloris Basin) is 1,300 kilometres in diameter. Mariner 10 photographed only part of the surface but there is no reason to doubt that the remainder is essentially similar.

Mercury has only an extremely tenuous atmosphere, corresponding to a laboratory vacuum. There is a relatively large heavy core and a weak magnetic field. Life there is certainly out of the question.

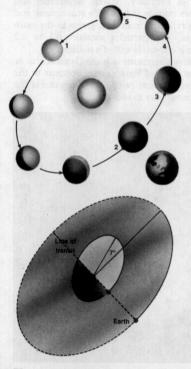

The phases of Mercury may be followed with a small telescope. When "new" (1) the planet is invisible (except during transit) and when "full" (2) it is on the far side of the Sun, so that it is best seen as a crescent (3), half (4) or gibbous (5). It is unwise to search for it with the Sun above the horizon, except with a telescope that is equipped with accurate setting circles.

Mercury passing directly between the Sun and the Earth is seen in transit and appears as a black disk against the brilliant solar surface; at these times it is obviously much darker than a sunspot. During transits Mercury cannot be seen with the naked eye and the best way to observe it is by projection with a telescope. The next transit is due on 13 November 1986.

50

This view of Mercury (*above*) was obtained with a 31.75 cm reflector. It shows only one faint shading and even this was at the limit of visibility. The coverage from Mariner 10 (*right*) shows a decidedly "lunar" landscape, with craters probably formed in the same way as those on the surface of the Moon.

The Caloris Basin is a ringed structure (there is every reason to suppose that the part of it unavailable to Mariner 10 is as regular as the part recorded) and the effects of its formation are widely shown over the surrounding regions. Fractures on the floor of the basin are much in evidence. Information, however, is still not complete because the angle of illumination over the basin was very low.

Day 32

Day 21

Day 42.5

Day 10.5

Day 58.6

Day 1

Mercury rotates slowly on its axis. It used to be thought that the rotation was synchronous—the same as that of the planet's "year" (88 Earth days)—in which case there would be an area of permanent daylight and an area of permanent night. In fact its true period is 58.6 days, two-thirds of a Mercurian "year". When best placed, as seen from Earth, the same face is always turned towards us.

Venus

Venus is the most brilliant of the planets, but through a telescope little can be seen apart from the characteristic phase; the surface is permanently hidden by layers of cloud.

Before the space age very little was known about the surface conditions on Venus, but in 1962 Mariner 2 by-passed the planet and showed that the surface temperature is very high—therefore disproving an older theory that there are oceans on the planet. Since then more probes have been to Venus; the soft-landers Veneras 9 and 10 even sent back one picture each before being put out of action by the intensely hostile conditions. The latest information comes from American and Russian orbiters which have been put into closed paths around Venus and have sent back details of the surface. It is now established that there is a vast rolling plain, together with two highland areas (Ishtar and Aphrodite) and high peaks, together with craters. At low levels, thunder and lightning is continuous; the temperature is around 480°C, the pressure of the atmosphere is at least 90 times that of the Earth's air and the clouds of Venus contain large quantities of sulphuric acid. It is clear that no Earth-type life can exist there.

When Venus is closest to the Earth (inferior conjunction) it is new, with the dark side turned towards Earth. As the phase increases, the diameter appears to shrink. When full (superior conjunction) Venus is on the far side of the Sun. Greatest brilliancy occurs when Venus is in the crescent stage where it is best seen with the naked eye.

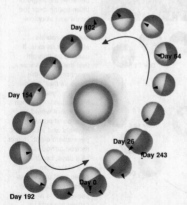

Day 102
Day 64
Day 154
Day 26
Day 243
Day 0
Day 192

Venus is unique in having an axial rotation period longer than the period of revolution around the Sun. The Venus "year" is 224.7 Earth days; the axial rotation period is 243 Earth days, but the rotation period of the upper clouds is only 4 days. To an observer on Venus, the interval between one sunrise and the next would be 118 Earth days. To complete the strangeness of the picture, Venus rotates in a retrograde or backward direction; east to west instead of west to east.

Photographs were obtained from the American Mariner 10 probe at seven-hour intervals in 1974 and confirmed the prediction that the rotation period is a mere four days for the upper clouds—though the solid planet has a rotation period of 243 days. The rotation is in the retrograde sense, so that the axial inclination is 178°—a case which is unique in the Solar System. In these pictures, one cloud region covering an area of approximately 1,000 square km is arrowed so that its shift due to the planet's rotation can be clearly seen.

Location		1985	1986	1987	1988
Jan.	1	Aqr.	Oph.	Lib.	Cap.
	16	Aqr.	Sgr.	Oph.	Aqr.
Feb.	1	Psc.	Cap.	Oph.	Aqr.
	16	Psc.	Aqr.	Sgr.	Psc.
Mar.	1	Psc.	Aqr.	Sgr.	Psc.
	16	Psc.	Psc.	Cap.	Ari.
Apr.	1	Psc.	Psc.	Aqr.	Tau.
	16	Psc.	Tau.	Aqr.	Tau.
May	1	Psc.	Tau.	Psc.	Tau.
	16	Psc.	Tau.	Ari.	Gem.
Jun.	1	Psc.	Gem.	Ari.	Tau.
	16	Ari.	Cnc.	Tau.	Tau.
Jul.	1	Tau.	Cnc.	Tau.	Tau.
	16	Tau.	Leo	Gem.	Tau.
Aug.	1	Ori.	Leo	Cnc.	Tau.
	16	Gem.	Vir.	Sex.	Gem.
Sep.	1	Cnc.	Vir.	Leo	Gem.
	16	Leo	Vir.	Leo	Cnc.
Oct.	1	Leo	Lib.	Vir.	Leo
	16	Vir.	Lib.	Vir.	Leo
Nov.	1	Vir.	Lib.	Lib.	Vir.
	16	Lib.	Vir.	Oph.	Vir.
Dec.	1	Lib.	Vir.	Sgr.	Lib.
	16	Oph.	Lib.	Sgr.	Lib.

The first picture to be transmitted from the surface of Venus was received in October 1975 from the Russian soft-landing probe Venera 9. Pictures showed a rock-strewn landscape with stones measuring up to a metre across. Wind velocities were low and the light level was compared to that of a cloudy winter's day in Moscow at noon.

53

Mars 1

Mars, the first planet beyond the Earth in the Solar System, is easy to recognize because of the strong red colour that led to it being named in honour of the God of War. At its brightest it outshines every other planet apart from Venus.

Mars in general is not easy to study with a small telescope but, under good conditions, adequate instruments show dark, well-defined features and white polar caps. The dark area were once thought to be old sea-beds filled with vegetation but this attractive idea has now been disproved and as yet there is no firm evidence of any life on Mars. Some of the dark regions, such as Syrtis Major, are elevated plateaux.

The Martian "day" is about half an hour longer than that of the Earth. The apparent drift of features across the disk is therefore obvious even over a short period of observation. Generally the atmosphere is transparent, except for occasional dust storms which may hide the surface features.

Location		1985	1986	1987	1988
Jan.	1	Aqr.	Vir.	Psc.	Lib.
	16	Aqr.	Lib.	Psc.	Oph.
Feb.	1	Psc.	Lib.	Psc.	Oph.
	16	Psc.	Oph.	Psc.	Oph.
Mar.	1	Psc.	Oph.	Ari.	Sgr.
	16	Ari.	Oph.	Ari.	Sgr.
Apr.	1	Ari.	Sgr.	Tau.	Sgr.
	16	Ari.	Sgr.	Tau.	Cap.
May	1	Tau.	Sgr.	Tau.	Cap.
	16	Tau.	Sgr.	Gem.	Cap.
Jun.	1	Tau.	Sgr.	Gem.	Aqr.
	16	Gem.	Sgr.	Gem.	Aqr.
Jul.	1	Gem.	Sgr.	Gem.	Psc.
	16	Gem.	Sgr.	Cnc.	Psc.
Aug.	1	Cnc.	Sgr.	Cnc.	Psc.
	16	Cnc.	Sgr.	Leo	Psc.
Sep.	1	Leo	Sgr.	Leo	Psc.
	16	Leo	Sgr.	Leo	Psc.
Oct.	1	Leo	Sgr.	Leo	Psc.
	16	Vir.	Cap.	Vir.	Psc.
Nov.	1	Vir.	Cap.	Vir.	Psc.
	16	Vir.	Cap.	Vir.	Psc.
Dec.	1	Vir.	Aqr.	Vir.	Psc.
	16	Vir.	Aqr.	Lib.	Psc.

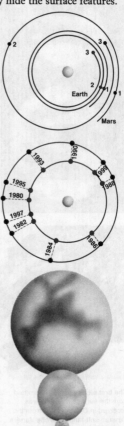

The revolution period of Mars is 687 days. The diagram (*top right*) shows the opposition of Earth and Mars (1). A year later the Earth has returned to position (1) but Mars has only travelled just over half its orbit (2). The Earth has to "catch it up" before the next opposition (3). The time between oppositions is roughly 780 days. Because the Martian orbit is relatively eccentric not all oppositions are equally favourable (*centre*). Opposition distance may vary from 56 million km (1986) to 101 million km (1980). The apparent diameter of Mars viewed from Earth (*right*) ranges between 25″.7 and 3″.5.

The white polar caps are the most striking features of the Martian surface and may be seen with a small telescope. The seasonal variations in size are easy to follow and record. In particular, the south polar cap is turned towards the Earth during the most favourable oppositions, making it ideal for observation. The diagram shows the shrinkage rate of the southern cap in 1956 as observed with a 25 cm reflector. The shrinkage was formerly attributed to the release of water vapour, which might support vegetation. Since the discoveries of the two Viking landers, vegetation is now considered unlikely although the polar caps do contain water ice.

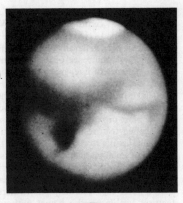

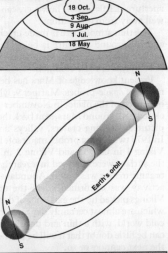

18 Oct.
3 Sep.
9 Aug.
1 Jul.
18 May

The axial inclination of Mars is 23° 58′, almost identical to that of the Earth's 23° 27′. The seasons are of the same type but much longer duration. There is, however, an important difference; like Earth, Mars is at its closest to the Sun during the summer in the Southern Hemisphere. Greater eccentricity of orbit means that the southern seasons are more extreme than the northern.

Earth's orbit

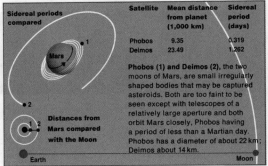

Sidereal periods compared

Satellite	Mean distance from planet (1,000 km)	Sidereal period (days)
Phobos	9.35	0.319
Deimos	23.49	1.262

Phobos (1) and Deimos (2), the two moons of Mars, are small irregularly shaped bodies that may be captured asteroids. Both are too faint to be seen except with telescopes of a relatively large aperture and both orbit Mars closely, Phobos having a period of less than a Martian day. Phobos has a diameter of about 22 km; Deimos about 14 km.

Distances from Mars compared with the Moon

Earth

Moon

Mars 2

The first reasonably accurate maps of Mars were drawn in the nineteenth century. In 1877 G. V. Schiaparelli, from Milan, compiled a chart and named the main features. These names have been generally retained, although recently modified in view of the space-probe discoveries. Schiaparelli also drew a network of fine, sharp lines which became known as canals and were regarded by some astronomers as being of artificial origin. It is now known that the canal network does not exist in any form and that it was purely illusory.

The most prominent dark markings are the Syrtis Major, a V-shaped feature close to the Martian equator, and the Acidalia Planitia in the north. Both of these are very easy to see with a small telescope when Mars is suitably placed near opposition. South of the Syrtis Major is Hellas, a circular basin which can at times be so bright that it is easy to confuse with a polar cap. The map shown here indicates the main features visible on the planet with telescopes of modest aperture. Little detail, however, can be seen when Mars is well away from opposition.

The thin atmosphere is made up chiefly of carbon dioxide and the ground pressure is below ten millibars, so that no surface water can exist. Clouds are frequent but extensive dust storms are less common.

Present knowledge of Mars has been obtained mainly by means of space-probes. Mariner 9, which reached the neighbourhood of the planet in November 1971 and was put into a closed path around Mars, sent back thousands of high-quality pictures, showing craters, valleys and towering volcanoes. In 1976 two Viking probes made soft landings on the surface, Viking 1 in Chryse and Viking 2 in Utopia. Samples from the surface were analysed in a search for life, but no trace of organic matter was found. Abundant evidence of past water activity was confirmed when the orbiting sections of the Vikings picked up the existence of deep canyons and features which are almost certainly dry river-beds. Though Mars is a cold world, with a thin and unbreathable atmosphere, there can be little doubt that it will be reached by astronauts within the next century or so.

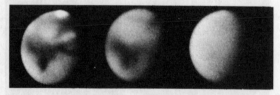

The Martian dust storm of 1971-72 was one of the greatest ever observed. The photographs were taken at the Lowell Observatory in Arizona. From left to right they show the beginning of the storm on 22 September 1971, the surface features partially obscured by 3 October and completely obscured by 21 October 1971. The surface had partly cleared by mid-December but was not fully exposed again until early 1972. Another storm occurred near the next opposition, in 1973.

South

North

Mars 3

Mare Chronium	
Electris	
Eridania	
Ausonia	
Hellas	
Hesperia Planum	
Mare Cimmerium	
Tyrrhena Planum	
Hadriaca Planitia	
Libya	
Isidis Planitia	
Syrtis Major	
Aethiopis	
Aeria	
Casius	
Elysium	
Utopia	
Trivium Charontis	

Mare Australe
Memnonia
Mare Sirenum
Propontis
Amazonis Planitia
Olympus Mons
Arsia Mons
Arcadia Planitia
Castorius Lacus
Solis Planum
Thaumasia
Tithonius Lacus
Tharsis
Tractus Albus
Argyre
Aurorae Planum
Xanthe
Chryse
Nilliacus Lacus
Acidalia Planitia
Mare Erythraeum
Sinus Margaritifer
Deucalionis Regio
Oxia Palus
Sinus Meridiani
Noachis
Pandorae Fretum
Sinus Sabaeus
Moab
Vastitas Borealis
Ismenius Lacus
Arabia
Hellespontus
Ausonia

57

Mars 3

Argyre Planitia is a relatively smooth plain that is sometimes visible with a small telescope, despite some variations in brightness. This Viking 1 photograph, taken from 19,000 km, shows an oblique view of the horizon. Argyre is surrounded by quite heavily cratered terrain, but it is impossible to see any craters with Earth-based telescopes.

Valles Marineris is an enormous canyon stretching nearly one-third of the way round Mars. It has a width of 402 km and reaches to a depth of 6.4 km in places. This section taken by Viking Orbiter 1 at a distance of 4,200 km illustrates a number of landslides on the far wall and deeply etched branch channels cutting into the plateau at the bottom.

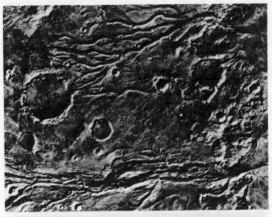

Chryse Planitia was the site selected for the first soft landing on Mars by the Viking Orbiter 1 probe. The surface mosaic was taken by the Orbiter at a distance of 1,680 km and covers an area of about 2,000 sq km. Numerous features are visible, including faults, lava flows, craterlets and several dry stream beds or channels.

The north polar cap, viewed from 4,000 km by Viking Orbiter 2. The detail shows white bands free of water ice, possibly the result of high-velocity winds.

The "sand-dune" crater Proctor, seen in the region of Hellespontus, photographed from Mariner 9. The dunes clearly indicate the direction of the prevailing winds.

Olympus Mons is the highest volcano on Mars and probably the highest anywhere in the Solar System. Its summit lies nearly 25 km above the general level of the surface and the multi-ringed caldera is 65 km in diameter. Olympus Mons is a typical shield volcano and is similar to those of the Hawaiian Islands on Earth, but on a much grander scale. It lies on the Tharsis ridge and is one of a group of gigantic volcanoes.

Asteroids

The minor planets, or asteroids, are dwarf worlds. Only one (Ceres) is as much as 1,000 kilometres in diameter and only Vesta is ever visible to the naked eye. Most of the asteroids move in the region between the orbits of Mars and Jupiter. It has been suggested that they are the fragments of a former planet that broke up, although most astronomers now believe that the asteroids never formed part of a larger body; their combined mass is much less than that of the Moon.

The first four asteroids (Ceres, Pallas, Juno and Vesta) were discovered between 1801 and 1807. Today several thousand asteroids are known but most are very faint. Some have orbits that swing them away from the main group. Thus Icarus (a mere 1 kilometre across) moves closer to the Sun than the orbit of Mercury, while the so-called Trojans move in the same orbit as Jupiter. Chiron, discovered in 1977, moves mainly between the orbits of Saturn and Uranus. Its exact nature is still unknown.

During a time exposure (*above*), a guided telescope will follow the sky's movement so that the stars appear as sharp points, but an asteroid will be moving with respect to the stars and so leave a trail. One trail is shown in this photograph by Max Wolf.

The main asteroid zone (*right*) lies between the orbits of Mars and Jupiter. A number of gaps in the zone are caused by the cumulative perturbations from Jupiter's pull of gravity. These are known as Kirkwood gaps, named after their discoverer.

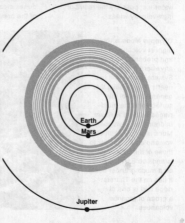

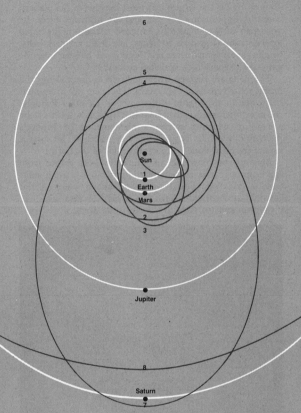

Asteroid	Distance (a.u.)	Period (years)	Inclination (degrees)	Diameter (km)
1 Icarus	1.08	1.12	23.0	1
2 Apollo	1.47	1.78	6.3	1
3 Hermes	1.64	2.10	6.2	0.5
4 *Melpomene	2.30	3.48	10.1	130
5 Ceres	2.77	4.60	10.6	1,003
6 Achilles	5.21	11.90	10.3	53
7 Hidalgo	5.82	14.04	42.5	16
8 Chiron	13.69	50.68	6.9	150–650?

*Melpomene has its own satellite

Of those asteroids outside the main zone, the Trojans (such as Achilles) are remote, moving in the same orbit as Jupiter. To date the most remote is Chiron, with an orbit lying between those of Saturn and Uranus. Hidalgo has an eccentric orbit that carries it from inside the path of Mars out almost as far as Saturn. Of the ''Earth-grazers'', Hermes came the closest in 1937 at a distance of only 770,000 km. Oddest of all the asteroids is Icarus. At aphelion it lies beyond Mars' orbit but at perihelion it is closer to the Sun than Mercury.

Jupiter 1

Jupiter, by far the largest and most massive of the planets, comes to opposition each year and is a brilliant object. Its yellowish disk is markedly flattened because of its rapid axial rotation (less than 10 hours) and there are dark belts, bright zones, spots, wisps and festoons to be seen over the gaseous surface, which is always changing.

The belts are regions where material is descending; the bright zones represent upcurrents. There are two prominent belts, one to either side of the equator. Of other features, the most interesting is the Great Red Spot, now known to be a kind of whirling storm—a phenomenon of Jupiter's meteorology, which is generally very violent. The surface area of the Spot alone is greater than that of the Earth.

Most of our knowledge of Jupiter has been gained from space-probes. It is thought that Jupiter has a small solid core, overlaid by liquid hydrogen, above which is the gaseous "atmosphere". A flat ring of particles also exists but is too thin and too faint to be seen from Earth.

Jupiter photographed from 28.4 million km by Voyager 1 (*above*) in 1979. The Great Red Spot has been observed for hundreds of years, yet never in the detail seen here. White ovals are also clearly visible. Those south of the Red Spot were seen to form about 40 years ago. The Earth-based photograph (*left*) was taken with a 39 cm reflector. The surface features are clearly defined. Note also the satellite Europa and its shadow in transit across the planet.

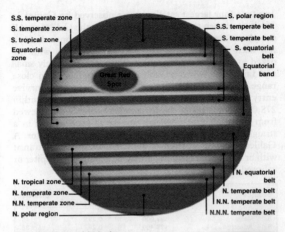

S.S. temperate zone
S. temperate zone
S. tropical zone
Equatorial zone

S. polar region
S.S. temperate belt
S. temperate belt
S. equatorial belt
Equatorial band

Great Red Spot

N. tropical zone
N. temperate zone
N.N. temperate zone
N. polar region

N. equatorial belt
N. temperate belt
N.N. temperate belt
N.N.N. temperate belt

The nomenclature of Jupiter's surface. The belts (dark areas) and the Red Spot do not change appreciably in latitude. Because Jupiter's rotation is so quick, the drift of the markings across the disk becomes quite obvious after only a few minutes' observation. The rotation period of the equatorial zone is about 5 minutes shorter than the rotation period of the rest of the planet; but various features, including the Red Spot, have periods of their own and therefore drift about in longitude over the surface.

Location		1985	1986	1987	1988
Jan.	1	Sgr.	Cap.	Aqr.	Psc.
	16	Sgr.	Cap.	Aqr.	Psc.
Feb.	1	Sgr.	Aqr.	Psc.	Psc.
	16	Cap.	Aqr.	Psc.	Psc.
Mar.	1	Cap.	Aqr.	Psc.	Psc.
	16	Cap.	Aqr.	Psc.	Ari.
Apr.	1	Cap.	Aqr.	Psc.	Ari.
	16	Cap.	Aqr.	Psc.	Ari.
May	1	Cap.	Aqr.	Psc.	Ari.
	16	Cap.	Aqr.	Psc.	Ari.
Jun.	1	Cap.	Aqr.	Psc.	Tau.
	16	Cap.	Psc.	Psc.	Tau.
Jul.	1	Cap.	Psc.	Psc.	Tau.
	16	Cap.	Psc.	Psc.	Tau.
Aug.	1	Cap.	Psc.	Psc.	Tau.
	16	Cap.	Psc.	Ari.	Tau.
Sep.	1	Cap.	Aqr.	Ari.	Tau.
	16	Cap.	Aqr.	Psc.	Tau.
Oct.	1	Cap.	Aqr.	Psc.	Tau.
	16	Cap.	Aqr.	Psc.	Tau.
Nov.	1	Cap.	Aqr.	Psc.	Tau.
	16	Cap.	Aqr.	Psc.	Tau.
Dec.	1	Cap.	Aqr.	Psc.	Tau.
	16	Cap.	Aqr.	Psc.	Tau.

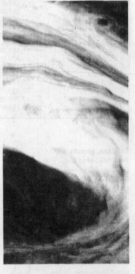

Jupiter is a world of violent activity, as the Voyager pictures show. The upper atmosphere is cold although near the core the temperature is at least as high as 30,000°C and may well be more. Hurricanes and cyclones swirl about in the upper gas; there are brilliant colours and cloud patterns are always changing. The Red Spot (the dark area in the photograph) is relatively calm but beyond its confines the scene is more violent.

Jupiter 2

Jupiter's four largest satellites were observed by Galileo in 1610; since then they have been known as the Galileans. All are planet-sized and all can be seen with any optical aid; a few keen-sighted people can even glimpse them with the naked eye. The remaining 12 satellites are extremely small and faint. In 1979 the Galileans were studied from close range by the Voyager probes, which managed to survive entry into Jupiter's powerful magnetic field to within 278,000 kilometres of the Jovian cloud tops. When viewed from Earth all four Galileans appear to keep almost in a straight line, as they lie in the plane of Jupiter's equator. A Galilean passing in front of Jupiter may appear in transit with or without its shadow; it may be occulted by Jupiter or eclipsed by its shadow.

Satellite		Mean distance from planet (km)	Sidereal period (days)	Diameter (km)
S.16	Metis	127,000	0.295	±40
S.14	Adastrea	128,400	0.297	±35
S.5	Amalthea	181,300	0.498	150×270
S.15	Thebe	225,000	0.678	±75
S.1	Io	421,600	1.769	3,632
S.2	Europa	670,900	3.551	3,126
S.3	Ganymede	1,070,000	7.155	5,276
S.4	Callisto	1,883,000	16.689	4,870
S.12	Leda	11,100,000	238.7	8
S.6	Himalia	11,470,000	250.6	170
S.10	Lysithea	11,710,000	259.2	19
S.7	Elara	11,743,000	259.7	80
S.12	Ananke	20,700,000	631	17
S.11	Carme	22,350,000	692	24
S.8	Pasiphaë	23,300,000	744	28
S.9	Sinope	23,700,000	758	21

Eclipses, transits and occultations of the Galilean satellites (*right*). A satellite can be eclipsed by Jupiter's shadow (1), can pass in front of Jupiter and appear in transit, together with its shadow (2), or be occulted by Jupiter (3). These phenomena are easily followed with a small telescope. Of the Galileans orbiting Jupiter (*lower right*). Amalthea is the closest, followed by Io, Europa, Ganymede and Callisto. The remaining outer satellites are asteroidal in nature; their orbits are strongly perturbed by the Sun and are not even approximately circular. The four outermost have retrograde motion.

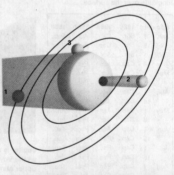

Io was photographed by Voyagers 1 and 2 and its surface is a strong red colour, attributed to sulphur. It may have a sulphur-rich surface overlying an ocean of liquid sulphur. Active volcanoes send up material that enters a torus around Jupiter. It seems certain that Io is associated with variations in the radio emissions received from Jupiter.

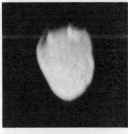

Amalthea is red and tiny and is an irregular body; Voyagers 1 and 2 showed only two of its surface features. It was discovered in 1892.

Europa, photographed by Voyager 2, has a very smooth surface that is virtually devoid of craters; it has very shallow "grooves" and is presumably icy.

Ganymede is also ice-surfaced; it has many craters, with strange "grooves" and one large, darker patch now named Galileo Regio.

Callisto, as seen from Voyager 2, is the outermost and darkest Galilean satellite. Its surface is icy and intensely cratered.

Saturn

Saturn is without doubt the most beautiful object in the sky. Like Jupiter, it has a gaseous surface; inside it is mainly liquid hydrogen and has a solid core. Belts are seen as well as occasional spots. The rings are visible in a small telescope: there are two bright rings (A and B) separated by the gap known as the Cassini Division and a semi-transparent ring (C) closer to the planet. Voyagers 1 and 2 (1980/81) by-passed Saturn and sent back images showing that the rings are much more complicated than had been thought. There are many minor divisions and some very faint rings. Of the satellites, Titan is the largest and has a dense atmosphere made up chiefly of nitrogen. The others are far smaller, mainly icy in composition and are cratered. The surface of Enceladus is in parts smooth and grooved or covered in small craters. Iapetus, Rhea, Tethys and Dione are fairly easy objects to observe. Iapetus has one bright and one dark hemisphere.

Amateur astronomer H. E. Dall's photograph of Saturn taken with a 39 cm reflector compares very well with the photograph (*opposite*) taken with a 2.5 m reflector. The ring system is well displayed and the Cassini Division, separating the two bright rings, is prominent.

Location		1985	1986	1987	1988
Jan.	1	Lib.	Sco.	Sco.	Sco.
	16	Lib.	Sco.	Sco.	Sco.
Feb.	1	Lib.	Sco.	Sco.	Sco.
	16	Lib.	Sco.	Sco.	Sgr.
Mar.	1	Lib.	Sco.	Sco.	Sgr.
	16	Lib.	Sco.	Sco.	Sgr.
Apr.	1	Lib.	Sco.	Sco.	Sgr.
	16	Lib.	Sco.	Sco.	Sgr.
May	1	Lib.	Sco.	Sco.	Sgr.
	16	Lib.	Sco.	Sco.	Sgr.
Jun.	1	Lib.	Sco.	Sco.	Sgr.
	16	Lib.	Sco.	Sco.	Sgr.
Jul.	1	Lib.	Sco.	Sco.	Sco.
	16	Lib.	Sco.	Sco.	Sco.
Aug.	1	Lib.	Sco.	Sco.	Sco.
	16	Lib.	Sco.	Sco.	Sco.
Sep.	1	Lib.	Sco.	Sco.	Sco.
	16	Lib.	Sco.	Sco.	Sco.
Oct.	1	Lib.	Sco.	Sco.	Sco.
	16	Lib.	Sco.	Sco.	Sco.
Nov.	1	Lib.	Sco.	Sco.	Sco.
	16	Lib.	Sco.	Sco.	Sgr.
Dec.	1	Sco.	Sco.	Sco.	Sgr.
	16	Sco.	Sco.	Sco.	Sgr.

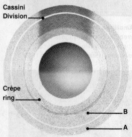

Cassini Division

Crêpe ring

B

A

This plan view of the ring system, drawn to scale, shows the rings to be perfectly circular. A and B are separated by the Cassini Division. Faint rings have been suspected.

The orbits of named satellites are shown to scale (*below*). The distances of the satellites range from 185,600 km (Mimas) to 12,960,000 km (Phoebe).

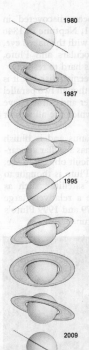

1980

1987

1995

2009

Saturn, taken from Hale Observatories, California, with a 2.5 m reflector, showing the rings at their widest opening. The ring system is extremely wide and very thin, measuring 272,300 km in diameter and only 16 km in depth.

Saturn's rings, photographed from Voyager 1 (*above*) on October 30, 1980, from a distance of 18 million km, were shown to have many "grooves". Narrow rings within the Cassini Division were also seen. Changing aspects of the rings are shown (*left*).

Satellite	Mean distance (km)	Period (days)	Diameter (km)
Atlas	137,670	0.60	60
1980 S27	139,353	0.61	140 × 80
1980 S26	141,700	0.63	110 × 70
Janus	151,422	0.69	220 × 160
Epimetheus	151,472	0.69	140 × 100
Mimas	185,600	0.94	390
Enceladus	238,200	1.37	500
Tethys	294,600	1.89	1,050
Telesto	294,600	1.89	50
Calypso	294,600	1.89	60
Dione	377,400	2.74	1,120
1980 S6	378,060	2.74	60
Rhea	526,800	4.52	1,530
Titan	1,200,000	15.95	5,150
Hyperion	1,482,000	21.28	410 × 220
Iapetus	3,558,000	79.33	1,440
Phoebe	12,960,000	550.4	±200

Uranus, Neptune, Pluto

The three outermost planets have been discovered in relatively modern times: Uranus in 1781, Neptune in 1846 and Pluto in 1930. Uranus is just visible with the naked eye, Neptune cannot be seen without binoculars and Pluto, being fainter than the 14th magnitude, is hard to resolve.

Telescopically Uranus appears as a greenish disk but it is too distant for surface detail to be seen, although two parallel equatorial belts similar to those of Jupiter and Saturn have been noted. Uranus has five satellites, which are all fainter than the 13th magnitude.

Neptune is slightly smaller than Uranus and is bluish rather than green; in small telescopes it has a stellar appearance. It has two satellites, which are difficult objects to see, being of the 13th and 18th magnitude. Pluto is thought to be similar in composition to an "icy satellite" such as Callisto. In 1978 it was found to have a relatively large satellite of its own, Charon. Between 1979 and 1999 Pluto's orbit will take it within the path of Neptune.

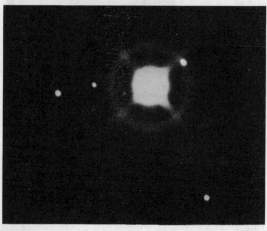

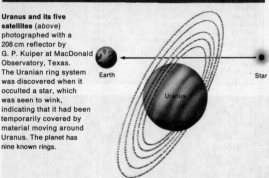

Uranus and its five satellites (*above*) photographed with a 208 cm reflector by G. P. Kuiper at MacDonald Observatory, Texas. The Uranian ring system was discovered when it occulted a star, which was seen to wink, indicating that it had been temporarily covered by material moving around Uranus. The planet has nine known rings.

Earth

Star

Uranus

Neptune and its two satellites
(*above*). The photograph shows
''spikes'' around the planet but these
are purely optical effects where
Neptune has had to be overexposed
to reveal the satellites Triton and
Nereid. Triton, partly obscured by

Neptune in the photograph, is a
massive body much larger than the
Moon and has a retrograde motion.
Nereid has a direct motion. It is
smaller than Triton with a diameter
of only 300 km; it has a highly
eccentric orbit.

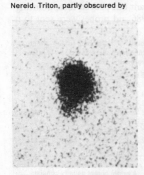

Pluto has an exceptional orbit that is
inclined 17° to the plane of the ecliptic
(*above*). The discovery in 1978 of the
satellite Charon, just visible as a blur
(*left*), gave a reliable estimate for the
mass of Pluto, which was found to be
only 0.14 per cent of that of the Moon.

Satellite	Mean distance from planet (1,000 km)	Sidereal period (days)	Diameter (km)
Uranus			
Miranda	130.5	1.414	300
Ariel	191.8	2.520	800
Umbriel	267.2	4.144	600
Titania	483.4	8.706	1,100
Oberon	586.3	13.463	1,000
Neptune			
Triton	353.0	5.877	3,700?
Nereid	5,560.0	359.881	300?
Pluto			
Charon	18.0	0.390	1,200?

Comets

A brilliant comet with a shining head and a tail stretching halfway across the sky is an awesome sight. Indeed, comets caused considerable alarm in ancient times as they were often regarded as harbingers of doom. Yet in comparison with a planet, a comet is flimsy and insubstantial: on several occasions the Earth has been known to pass unharmed through a comet's tail.

Comets have been aptly described as "dirty ice-balls". The main nucleus is composed of rocky fragments held together with ice such as frozen methane, ammonia, carbon dioxide and water. Surrounding the nucleus is the coma or head, which is made up of dust and gas. The Sun makes the comet visible by floodlighting the coma. When the comet nears the Sun, the radiation pressure of the sunlight, and the solar wind, drive the dust and gases away from the coma. This produces a tail that usually consists of both dust and gas. However, many faint comets do not develop a tail.

A number of comets move round the Sun in fairly small generally elliptical orbits that last a few years; others have longer cycles. However, most really brilliant comets (except Halley's) have orbital periods of thousands or even millions of years, which means that they cannot be predicted. Few have been seen in the twentieth century. Kohoutek's comet of 1973 was expected to be brilliant but proved disappointing. Halley's Comet returned in 1986 and probes were sent to it.

Comet	Period (years)	Orbital inclination (degrees)	Distance from Sun (in a.u.)		Next return
			min.	max.	
Brooks 2	6.9	5.6	1.8	5.4	1987
Encke	3.3	11.9	0.3	4.1	1987
Tuttle	13.8	54.7	1.0	10.3	1995
Finlay	6.9	3.4	1.1	6.2	1988
D'Arrest	6.2	18.0	1.4	4.7	1989
Crommelin	27.9	28.9	0.7	18.0	2012
Giacobini-Zinner	6.5	31.7	1.0	6.0	1991
Halley	76.1	162.2	0.6	18.0	2061

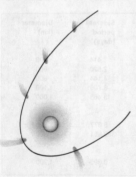

A bright comet usually consists of three principal parts: the compact nucleus, the surrounding gas and dust cloud (the coma) and the tenuous tail. Most comets move in highly elliptical orbits and only become conspicuous when near perihelion. At this point one or more tails begin to develop (*left*). The tail always points more or less directly away from the Sun. As the comet recedes the tail declines. There are two main types of tail: the first is usually straight and made up of ionized molecules driven from the coma by the solar wind; the second is curved and made up of dust driven out by solar radiation pressure.

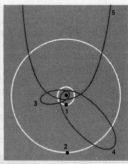

Typical comet orbits (*above*) are compared with the orbits of Jupiter (1) and Neptune (2). The smallest orbit (3) is that of a short-period comet, which has a period of a few years. Halley's comet (*left*) is the only conspicuous member of the long-period class (4). Comets with very long periods (5) have highly eccentric orbits that are almost parabolic. Kohoutek, photographed from Skylab (*below*), is an example. Arend-Roland (*bottom*) is another even more spectacular example.

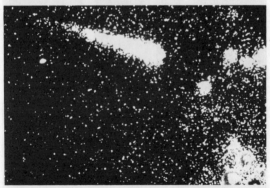

Meteors

Meteors may be regarded as the debris of the Solar System. They are too small and too fragile to reach the ground intact. When they dash down into the upper air, moving at speeds of up to 45 kilometres per second, they become intensely hot and destroy themselves in the streaks of luminosity which we call shooting stars.

Many meteors move around the Sun in swarms; each time the Earth passes through a swarm we see a shower of shooting stars. The most spectacular annual shower is known as the Perseids (25 July – 18 August). In addition to meteor showers, there are also non-shower or sporadic meteors, which may come from any direction at any moment.

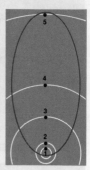

The Leonid meteor shower (*above* and *left*) intersects the orbit of the Earth (1), passing Mars (2), Jupiter (3), Saturn (4) and Uranus (5). It emanates from a point called the radiant.

Meteor shower	Date of return	Maximum	Location	Comment
Quantrantids	1-6 Jan.	4 Jan.	Boo.	Quite fast, blue
Corona Australids	14-18 Mar.	16 Mar.	CrA.	
Lyrids (April)	19-24 Apr.	21 Apr.	Lyr.	Fast, brilliant
η -Aquarids	1-8 May	5 May	Aqr.	Fast, persistent
Lyrids (June)	10-21 Jun.	15 Jun.	Lyr.	Blue
Ophiuchids	17-26 Jun.	20 Jun.	Oph.	
Capricornids	10 Jul.-15 Aug.	25 Jul.	Cap.	Yellow, very slow
δ -Aquarids	15 Jul.-15 Aug.	28 Jul.	Aqr.	Slow, long paths
Pisces Australids	15 Jul.-20 Aug.	30 Jul.	PsA.	
Capricornids	15 Jul.-25 Aug.	1 Aug.	Cap.	Yellow
ι -Aquarids	15 Jul.-25 Aug.	6 Aug.	Aqr.	
Perseids	25 Jul.-18 Aug.	12 Aug.	Per.	Fast, fragmenting
κ -Cygnids	18-22 Aug.	20 Aug.	Cyg.	Bright, exploding
Orionids	16-26 Oct.	21 Oct.	Ori.	Fast, persistent
Taurids	10 Oct.-30 Nov.	1 Nov.	Tau.	Slow, brilliant
Leonids	15-19 Nov.	17 Nov.	Leo.	Fast, persistent
Phoenicids	—	4 Dec.	Phe.	
Geminids	7-15 Dec.	14 Dec.	Gem.	White
Ursids	17-24 Dec.	22 Dec.	UMi.	

Meteorites

In the same way as meteors are associated with dead comets so meteorites, which are much larger lumps of rock, are thought to be associated with fragmented asteroids. They are of two main types: irons (siderites) and stones (aerolites), although there is no hard-and-fast distinction and many intermediate types are known.

Most museums have meteorite collections. The largest specimen on public view is the Ahnighito (Tent), which weighs more than 30 tonnes. It was found in Greenland by Robert Peary and is now in the Hayden Planetarium, New York. This is dwarfed by the Hoba West Meteorite, still lying where it fell at Grootfontein, Namibia, in prehistoric times. It weighs at least 60 tonnes. In recent years many meteorites have been discovered in Antarctica, where they have lain undisturbed for long periods; but whether the Hoba West record will be equalled remains to be seen. There have been many recorded meteorite falls. The last in Britain were those of 24 December 1965, when a meteorite broke up during descent and scattered fragments round the village of Barwell in Leicestershire; and of 25 April 1969, when the main mass fell into the Irish Sea.

Major falls are rare and there is no reliable record of any human casualty caused by a meteorite. It was formerly believed that meteoritic bodies would prove a major hazard in space research but it is now clear that the danger is less than was originally thought.

The damage to the pine forest in Siberia in 1908 (*left*) was thought to be caused by a meteorite. No fragments were ever recovered however. It is likely that the object was part of a small comet's nucleus. The meteorite crater in Arizona (*below*) was caused by an impact in prehistoric times. It is over 1 km wide.

Stars and their evolution

The stars are suns, but they are certainly not alike, and there is a tremendous range in size, luminosity and in surface temperature. The hottest stars have temperatures of at least 80,000 degrees centigrade, the coolest, below 3,000 degrees. These differences are shown by differences in colour. The spectra are also extremely diverse and the stars are divided into certain classes, each denoted by a letter of the alphabet. Stars of types O, B and A are very hot, and white or bluish-white; types F and G are yellow; K, orange; and M, R, N and S, orange-red.

According to modern theory, a star begins its career by condensing out of the material in a nebula, that is to say, a cloud of dust and gas. Hydrogen, the most plentiful element in the universe, makes up much of this gas and is the essential stellar "fuel"; but the whole sequence depends upon the initial mass of the star.

A star of mass less than 0.1 that of the Sun will shrink, because of the effect of gravitation, but will never become hot enough for nuclear reactions to begin. It will simply fade, becoming a dim, red star before turning cold and dead.

A star of between 0.1 and 1.4 solar mass will behave very differently. When the core temperature has reached about ten million degrees centigrade, nuclear reactions will begin; hydrogen will be converted to helium, and the star will settle down to a long period of stable existence. Eventually there will be no more available hydrogen. The core will shrink and heat up once more; different nuclear reactions will begin and the star's outer layers will expand and cool, so that the star will become a red giant. (The Sun will enter the red-giant stage in about five to seven thousand million years from now.) When all the nuclear energy is exhausted, the star will collapse into a small, super-dense object known as a white dwarf. The various parts of the broken-up atoms will be crammed together and the density may reach a million times that of water.

A star that is more massive still will run through its lifespan much more quickly and will die in a more spectacular manner. There will be a tremendous outburst, known as a supernova explosion, and much of the star's material will be blown away into space, leaving an expanding gas cloud together with a stellar remnant made up of neutrons—atomic particles with no electrical charge. A cupful of neutron-star material would weigh thousands of millions of tonnes. When a neutron star spins and sends out rapidly varying radio emissions, it is known as a pulsar.

Finally, there are stars of even greater mass. Once the final collapse starts, it is so rapid and so violent that nothing can stop it. The end product is a stellar remnant so small and pulling so strongly that not even light can escape from it, and the old star is surrounded by an area that is virtually cut off from the rest of the universe—making what astronomers have come to call a black hole.

Supergiant: Antares, the core temperature may be 100 million degrees.

Red giant: Arcturus, an orange star 113 times as luminous as the Sun.

Red giant

Yellow dwarf: the Sun, at least 5,000 million years old.

Yellow dwarf

White dwarf: Sirius B, with a diameter less than three times that of the Earth.

White dwarf

Neutron star: the "powerhouse" contained by the Crab Nebula.

Black hole: Cygnus X-1, a hot supergiant has a suspected black-hole companion.

Our Galaxy

Our Galaxy, or star system of which the Sun is a member, contains about 100,000 million stars, clusters, nebulae and much interstellar dust and gas. It has an overall diameter of perhaps 100,000 light-years (although some astronomers believe this to be an over-estimate) and the central "bulge" has a thickness of about 20,000 light-years.

The Galaxy is a flattened system. It has been compared, graphically if unromantically, to two fried eggs clapped together back to back. It is this shape that gives rise to the Milky Way appearance—that glorious band of radiance stretching right round the sky and the subject of innumerable legends. Any telescope, or even binoculars, will show that the Milky Way is made up of stars and at first glance it might seem that they are crowded together. Once again, appearances are deceptive. An observer looking along the main plane of the Galaxy sees many stars in almost the same direction and this accounts for the Milky Way. It is nothing more than a line-of-sight effect. Dark "rifts" may be seen here and there in the Milky Way band, indicating dark, obscuring material.

Seen in plan, the Galaxy appears as a spiral, rather like a huge Catherine-wheel. This is not surprising because many other galaxies are spiral in form. It also rotates. The Sun, at its distance of some 32,000 light-years from the centre of the system, takes 225 million years to complete one revolution—a period sometimes nicknamed the "cosmic year". Unfortunately it is impossible to see right through to the centre of the Galaxy because there is too much obscuring matter in the way, but radio waves can pass through this material. The galactic centre lies beyond the lovely star clouds seen in the constellation of Sagittarius.

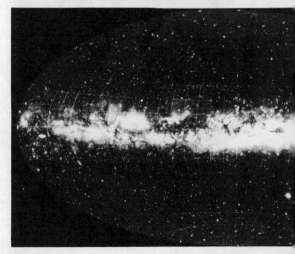

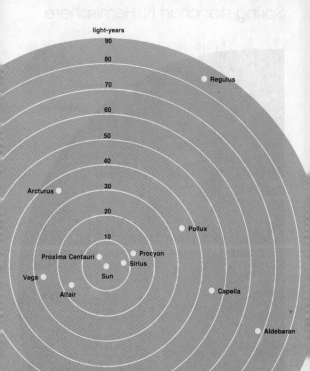

light-years

90
80
70 ● Regulus
60
50
40
30
● Arcturus
20
10 ● Pollux
● Proxima Centauri ● Procyon
Vega ● ● Sirius
● Sun
Altair ● ● Capella
● Aldebaran

Proxima Centauri is the nearest known star to Earth. It is a dim red dwarf 4.2 light-years away, although it is too faint to be seen with the naked eye. Other more distant stars, such as Sirius and Arcturus, are much brighter and have magnitudes of −1.4 and −0.1 respectively.

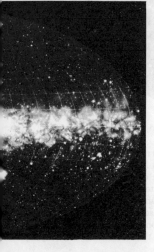

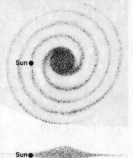

Sun ●

Sun ●

(Above) **The true shape of the Galaxy.** In plan, it is a spiral; from the "side", it would appear flattish with a central bulge.

(Left) **Mosaic of the Milky Way** as photographed in Sweden from the Lund Observatory.

77

Spring star chart N. Hemisphere

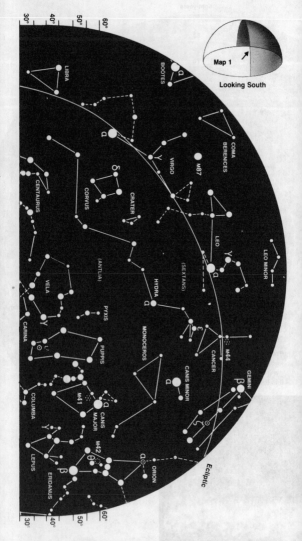

Orion is on view although it is now descending in the south-west and Sirius (Alpha Canis Majoris) is prominent. High in the south is the distinctive constellation of Leo with its leader Regulus (Alpha). Castor (Alpha Geminorum) and Pollux (Beta) are also high and Virgo is gaining altitude in the south-east. Much of the southern aspect is occupied by Hydra; this is the largest constellation in the sky but it contains only one bright star, the reddish Alphard (Alpha).

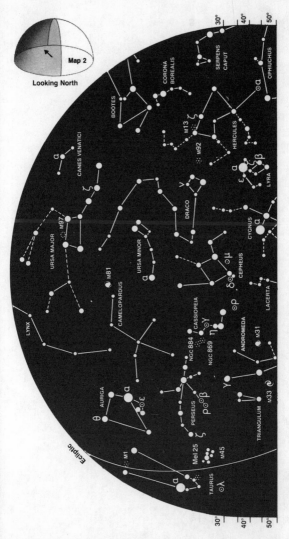

Ursa Major, the Great Bear, is now near the zenith and acts as a useful pointer to other groups. The "tail" of the Bear, for example, shows the way to Arcturus (Alpha Boötis). The W shape of Cassiopeia is relatively low, although from Britain and the northern United States it is circumpolar. Capella (Alpha Aurigae) is high; therefore Vega, on the opposite side of the pole, is low down and is so near the horizon that it may not be seen at all.

79

Summer star chart N. Hemisphere

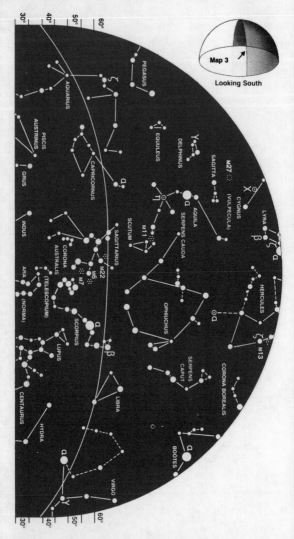

Map 3
Looking South

The sky during summer evenings is dominated by the so-called Summer Triangle of alpha stars: Vega in Lyra, which is almost overhead and has a bluish hue; Altair in Aquila, with its fainter star to either side; and Deneb in Cygnus, which is shown here on the northern aspect chart. Low down Antares (Alpha Scorpii) in Scorpio is at its best; it is the reddest of the brilliant stars. The area of Sagittarius is very rich, with magnificent star clouds.

80

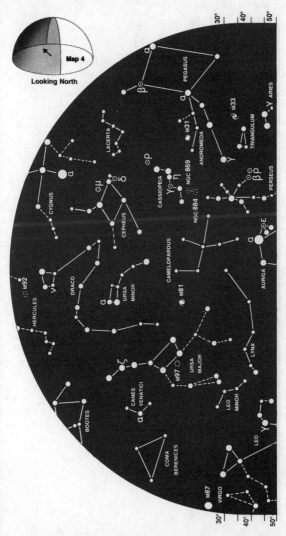

Looking North
Map 4

The northern aspect appears somewhat barren compared to the south. Ursa Major remains at a respectable altitude and Cassiopeia is also easy to see. The Square of Pegasus is coming into view in the east and Leo has almost disappeared. Capella (Alpha Aurigae) skirts the horizon, although from northern Europe and America it never actually sets. Much of the northern aspect is occupied by large, faint constellations such as Camelopardus, Lynx and Draco.

81

Autumn star chart N. Hemisphere

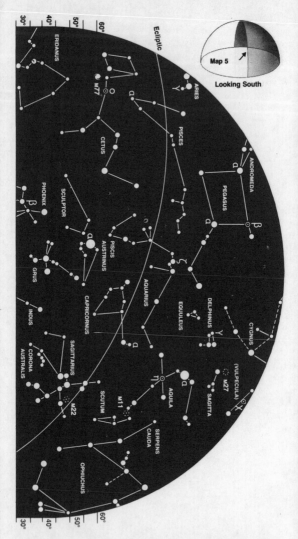

The main autumn constellation is Pegasus, which has four main stars making up a well-defined square, even though the brightest of the four, Alpheratz (Alpha) is illogically transferred to the adjacent constellation of Andromeda. Below the Square of Pegasus and not far above the horizon, when viewed from northern Europe and America, lies Fomalhaut (Alpha Piscis Austrini) of the 1st magnitude. Altair (Alpha Aquilae) remains high.

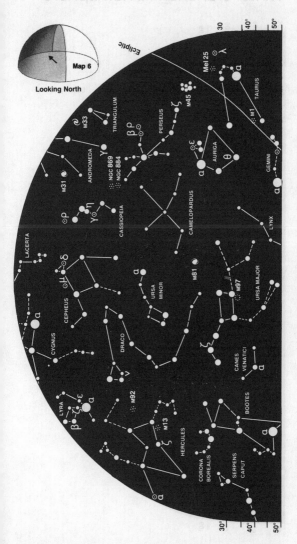

Looking North

Map 6

Vega (Alpha Lyrae) and Deneb (Alpha Cygni), in the Summer Triangle, are prominent and the cross of Cygnus is almost overhead. This whole area is very rich and well worth sweeping with binoculars. Ursa Major is at its lowest; Capella (Alpha Aurigae) is conspicuous in the north-east and Cassiopeia and Perseus are relatively high up. This is the best time of the year for seeing M31, the Great Spiral in Andromeda.

83

Winter star chart N. Hemisphere

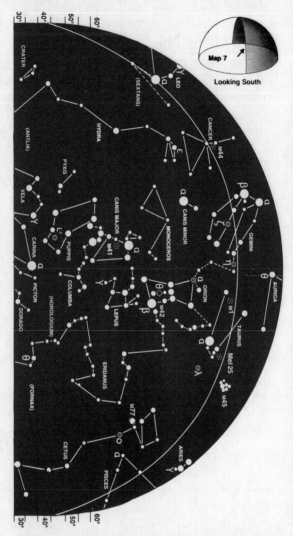

Map 7

Looking South

The southern aspect during winter is dominated by Orion and its two most brilliant stars, Rigel (Beta Orionis) and Betelgeuse (Alpha Orionis). Sirius (Alpha Canis Majoris) is at its best and is twinkling strongly; also in the Orion area are Gemini, Taurus (with the splendid clusters, Hyades and Pleiades) and Procyon (Alpha Canis Minoris). Capella (Alpha Aurigae) is almost overhead. The western area is occupied by the large, dim constellation Cetus.

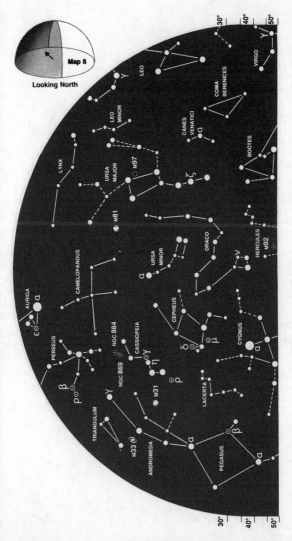

Map 8

Looking North

The northern aspect shows Ursa Major to the east, almost pointing down to the horizon. Its two pointers show the way to Ursa Minor, which contains Polaris, the Pole Star. Cassiopeia is still high in the north-west and takes the form of a W or an M of stars. The Milky Way flows through Cassiopeia, so that the whole area is rich in faint stars. Leo is rising in the east, Deneb is low down and Vega is so near the northern horizon that it is unlikely to be seen.

85

Spring star chart S. Hemisphere

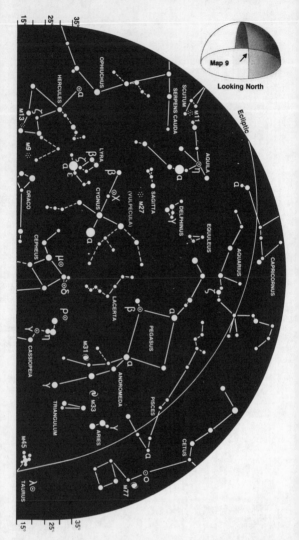

The triangle made up of Vega (Alpha Lyrae), Altair (Alpha Aquilae) and Deneb (Alpha Cygni) is just evident. High in the north the Square of Pegasus is prominent and from it leads off the line of stars marking Andromeda. Countries like South Africa and most of Australia never see the Great Spiral in Andromeda high in the sky. This is a good time of the year for seeing the richest part of the northern Milky Way, in the regions of Cygnus and Aquila.

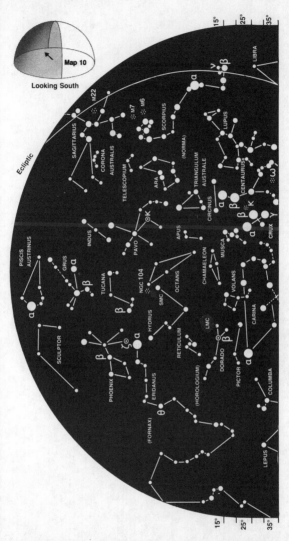

Map 10
Looking South

Fomalhaut, 1st-magnitude Alpha Piscis Austrini (also known as Piscis Australis), is almost overhead and its isolated position makes it easy to identify. Other constellations high in the sky are the Southern Birds, Grus, Pavo, Tucana and Phoenix. This is a rather confusing area; only Grus is distinctive, with its two 2nd-magnitude stars, Alpha and Beta. Scorpio is descending in the west. Crux, Centaurus and Canopus are all low in the south.

87

Summer star chart S. Hemisphere

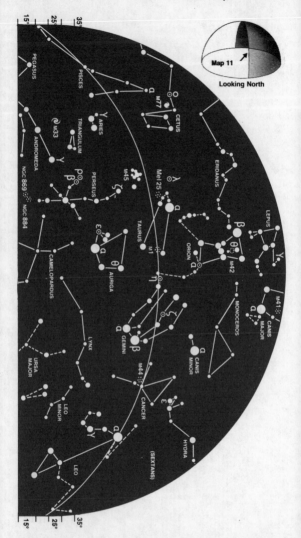

The northern aspect is dominated by Orion. The three stars of the Belt point upwards to Sirius (Alpha Canis Majoris), which is just at the zenith and therefore twinkles less than when seen from northern Europe and America, where it is always low down. Leo is on view and Castor and Pollux (Alpha and Beta Geminorum) are high. Auriga, led by Capella (Alpha), is reasonably high in the north. Perseus can also be seen although not to advantage.

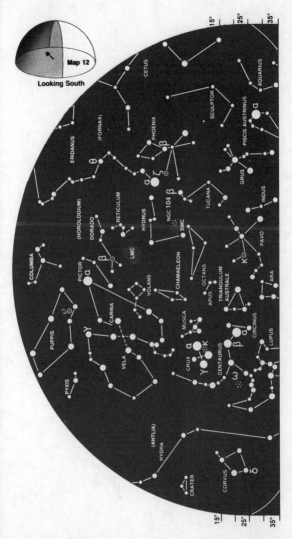

Map 12
Looking South

Many of the richest of the southern constellations are seen at their best during summer evenings. Canopus (Alpha Carinae), outshone only by Sirius, is high; the area is crowded with magnificent star fields and is crossed by the Milky Way. Crux Australis is well above the horizon near the two pointers, Alpha and Beta Centauri; on the far side of the pole Achernar (Alpha Eridani) stands out. The Magellanic Clouds in the south polar area are also high.

Autumn star chart S. Hemisphere

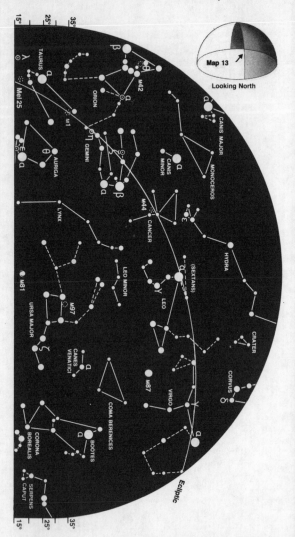

Map 13

Looking North

Orion is still above the horizon and Sirius (Alpha Canis Majoris) is high in the west. The northern part is dominated by Leo and this is a good time for seeing Praesepe, M44, in Cancer, which lies roughly midway between Leo and Gemini. There are no brilliant stars overhead, since the area of the zenith is largely occupied by the immense, dull Hydra. This is, however, the best time of the year to see Ursa Major, which is very low in the far north of the sky.

90

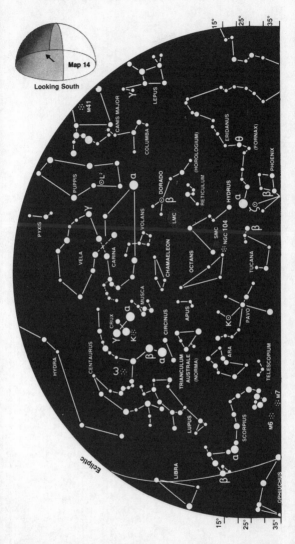

The Milky Way reaches right across the southern aspect and there are endless rich star fields. Crux Australis is now high. The brightest globular cluster in the sky is Omega Centauri, which is a hazy patch clearly visible with the naked eye. Canopus (Alpha Carinae) and the Magellanic Clouds are high up. Eridanus, with its faint stars, occupies the western region; Achernar (Alpha) is not far above the horizon and 47 Tucanae (NGC 104) is a prominent globular cluster.

91

Winter star chart S. Hemisphere

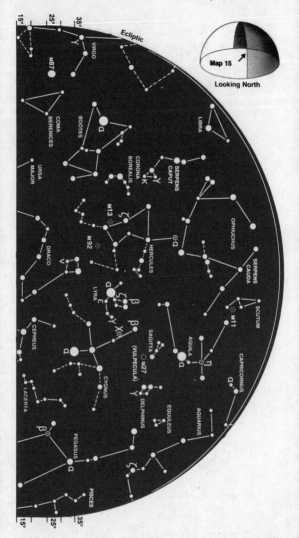

Looking North

Map 15

Three bright northern stars are on view, Vega (Alpha Lyrae), Altair (Alpha Aquilae) and Deneb (Alpha Cygni), make up the so-called Summer Triangle. Vega, with its pronounced bluish colour, is easy to find. The orange Arcturus (Alpha Boötis), and the prominent arc of stars marking Corona Borealis are distinctive. However, much of the northern aspect is occupied by the large, dim groups of Hercules, Ophiuchus and Serpens.

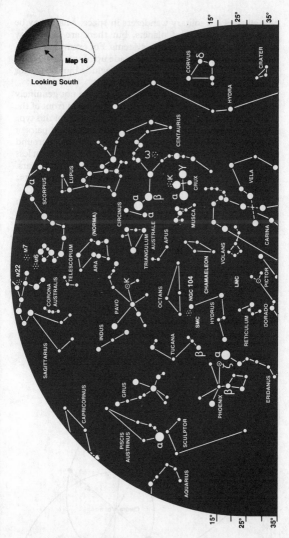

Map 16

Looking South

Scorpius is one of the most magnificent constellations in the sky and it is at its best during winter evenings when not far from the zenith. The redness of Antares (Alpha) stands out and this long line of bright stars conjures up an idea of some celestial insect. Fomalhaut (Alpha Piscis Austrini) is prominent in the south-east and Crux Australis is still high. Canopus (Alpha Carinae), however, is more or less out of view. Centaurus is dominant.

93

Double stars

Not all stars are solitary wanderers in space. Many may be attended by families of planets, but there are also many double stars and even multiple systems. For example, Castor, Alpha Geminorum, is a multiple made up of two stars. Each component is again double, and also included in the Castor system is a separate pair of faint red stars.

In some cases the two stars of a couple are not genuinely associated; one component merely lies almost in front of the other. Rather surprisingly, however, optical pairs of this type are not nearly so numerous as physically associated pairs of binary systems. In a binary, the two components move around their common centre of gravity. Periods may range from less than half an hour for ultra-close pairs to millions of years. For example, Arich, in Virgo, has almost equal components; the revolution period is 180 years.

Some doubles are separable with the naked eye. Mizar, in Ursa Major, has a naked-eye companion (Alcor), and in a small telescope Mizar itself is seen to be double. Other wide, easy pairs are Alpha Centauri and Alpha Crucis, two of the most brilliant stars of the Southern Hemisphere. Sometimes the components are quite unequal; thus the brilliant Sirius Alpha Canis Majoris has a dim white companion with only one ten-thousandth of the luminosity of Sirius itself.

It used to be thought that a binary pair resulted from the break-up of a formerly single star, but it is now thought more likely that the components of a binary were formed from the same cloud of material in the same region of space.

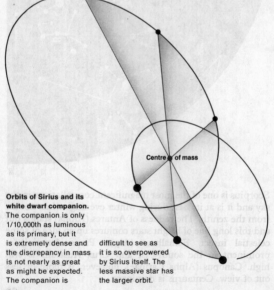

Orbits of Sirius and its white dwarf companion.
The companion is only 1/10,000th as luminous as its primary, but it is extremely dense and the discrepancy in mass is not nearly as great as might be expected. The companion is difficult to see as it is so overpowered by Sirius itself. The less massive star has the larger orbit.

Variable stars

Variable stars, as their name implies, brighten and fade over relatively short periods. Yet not all fluctuating stars are truly variable. With eclipsing binaries, such as Algol in Perseus, the changes result from one component passing in front of the other and cutting out some or all of its light. Algol "winks" regularly every two and a half days, falling from magnitude 2 to below 3.

With a true variable star, the changes are intrinsic. Of special importance are the Cepheids, so called because the most celebrated example is Delta Cephei in the northern sky. The variations are as regular as clockwork: Delta Cephei takes 5.37 days to pass from one maximum to the next. It has been found that the real luminosity of a Cepheid is linked with its period. As a result, observation of the period reveals the real luminosity and the distance may be worked out. Cepheids are invaluable "standard candles" in space.

There are also many variables that have longer periods, mainly red giants. The best known of these is Mira in Cetus. Here both the periods and amplitudes are irregular to some extent. True irregular variables are unpredictable and some are unstable and explosive. Even more spectacular are novae that suddenly flare up to many times their usual brilliance, remaining bright for a few days, weeks or months before fading back to obscurity. The brightest nova of modern times, seen in Aquila in 1918, surpassed every star in the sky apart from Sirius, but it soon declined and has now become very faint.

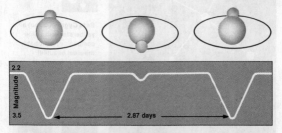

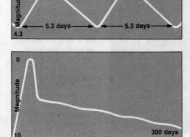

Theory of an eclipsing binary. Algol, Beta Persei (*above*), is an eclipsing binary. Where the larger, fainter component passes in front of the brighter, there is a marked drop in light; when the fainter is eclipsed, the fall in magnitude is slight. The light-curve of Delta Cephei (*centre*) has regular fluctuations but the light-curve of Nova Aquilae 1918 (*left*) shows an abrupt rise and a gradual fall.

Double stars N. Hemisphere

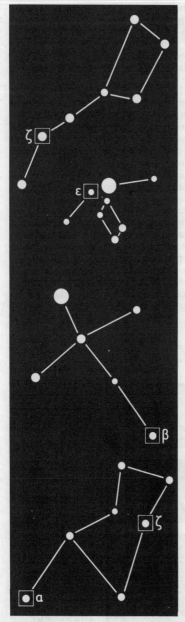

Zeta Ursae Majoris (Mizar) makes up a naked-eye pair with Alcor. With a small telescope Mizar is itself double; the components are rather unequal (magnitude 2.4, 3.9) and separated by 14″.5. *See maps 4, 6, 8, 13.*

Epsilon Lyrae, near Vega, is a fine example of a multiple star. The two main components (separation 207″.8) are visible to the naked eye. In an 8 cm telescope each component is again double. *See maps 2, 6, 9, 15.*

Beta Cygni (Albireo) a yellow star with a blue companion, is probably the most beautiful double in the sky. It can easily be seen at the foot of Cygnus with a small telescope or strong binoculars. *See maps 9, 15.*

Zeta Herculis is a binary with a 34 year period. The magnitudes are 3.1 and 5.6, but the separation is always below 2″, so this is not an easy object for a small telescope. Alpha Herculis (Rasalgethi) is a red supergiant with a 5.4 magnitude companion at 4″. 6, appearing greenish. *See maps 2, 6, 9, 15.*

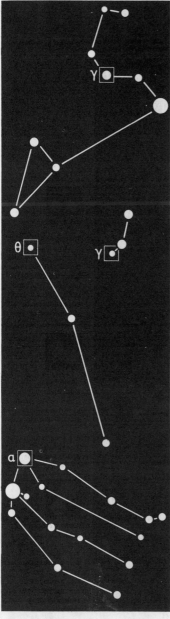

Gamma Leonis
(Algieba), an easy
double (separation over
4″), is a binary with a
period of 407 years.
The primary is of
magnitude 2.6, with
a 3.8 magnitude
companion. *See
maps* 1, 4, 11, 13.

Gamma Arietis
(Mesartim) is a wide,
easy double. Its
components are
equal, and the
separation is over 8″.
This is an optical pair,
not a binary. Both stars
have the same spectral
type. *See maps* 4, 9, 11.

Theta Serpentis (Alya)
is a particularly easy
double; the components
are exactly equal and
are separated by 22″.6
—a true pair of stellar
twins. Like Gamma
Arietis both stars have
the same spectral type.
See maps 3, 5, 15.

Alpha Geminorum
(Castor) has two
bright components, at
present 1″.8 apart, with a
period of 350 years.
Both components are
spectroscopic binaries
and have magnitudes of
2.0 and 2.9. Formerly
very wide and easy,
Castor has now closed
up, although it is not
difficult to separate.
See maps 7, 11, 13.

97

Double stars S. Hemisphere

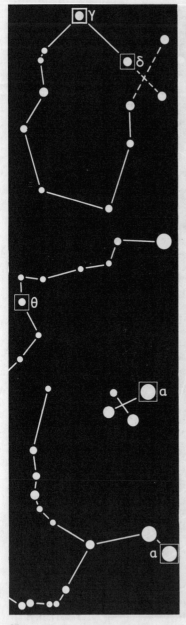

Gamma Velorum at 41″ is an easy double. Gamma[1] is magnitude 2.2 with Gamma[2], magnitude 4.8. Each component is itself a spectroscopic binary. They are about 520 light-years away. Delta[1] Velorum is magnitude 2.0 with a companion, magnitude 6.5. It has a separation of 2″.9. *See maps 1, 12, 14.*

Theta Eridani, at 8″.5, is a fine double that is separable with a small telescope. The combined magnitude is 2.9; the individual magnitudes are 3.4 and 4.4. They are at a distance of 135 light-years. *See maps 7, 10, 14.*

Alpha Crucis has a separation of 4″.7 and it can be easily separated with a small telescope. The individual magnitudes are 1.4 and 1.9 giving a combined magnitude of 0.8. *See maps 12, 14, 16.*

Alpha Centauri is a superb binary; the individual magnitudes are 0.0 and 1.7 giving a combined magnitude of −0.27. It has a separation of 22″; the revolution period is only 80 years. *See maps 12, 14.*

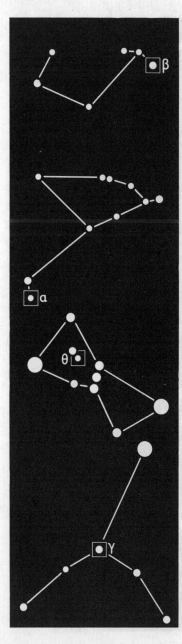

Beta Tucanae is a complicated system. Beta[1] and Beta[2] have a separation of 27".1 and magnitudes of 4.5. Beta[1] has a 14th-magnitude companion; Beta[2] is a close binary (magnitudes 4.9, 5.7). *See maps 10, 12, 14, 16.*

Alpha Capricorni is a naked-eye pair at 376". Alpha[1] (magnitude 4.5) has an optical 9th-magnitude companion. Alpha[2] (magnitude 3.8) has a companion with a magnitude of 10.6. *See maps 3, 5.*

Theta Orionis is the most famous multiple system, known as the Trapezium. The four main stars have magnitudes of 6.0, 7.0, 7.0, 7.5. The system is immersed in the Great Nebula, M42. *See maps 7, 11, 13.*

Gamma Virginis has equal components and magnitudes of 3.6 and 3.6, giving a combined magnitude of 2.8. It is one of the best doubles for small telescopes. The period is 180 years. Formerly it was wide and very easy; it has now closed to below 5" and will soon be difficult to separate. *See maps 1, 13.*

Variable stars N. Hemisphere

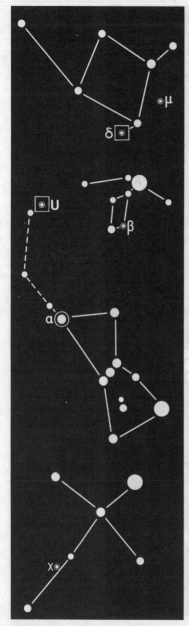

Delta Cephei is the prototype Cepheid and varies between magnitudes 3.5 and 4.4 in a period of 5.37 days. Mu Cephei varies between magnitude 3.6 and 5.1. It is probably the reddest of the naked-eye stars. *See maps* 2, 6, 8, 9.

Beta Lyrae appears variable but is in fact an eclipsing binary. The period is 12.9 days and the magnitude ranges from 3.4 to 4.3; there are alternate deep and shallow minima. The components are almost touching each other. *See maps* 2, 6, 9, 15.

Alpha Orionis, known as Betelgeuse, is a semi-regular variable with a period of about 2,070 days. The extreme range is probably from magnitude 0.2 to 0.9. U Orionis varies from magnitude 5.5 to 12.6; it has a 372-day period. *See maps* 7, 11, 13.

Chi Cygni is a Mira-type variable; with a period of 407 days, it has a very great range. The brightest maxima are of magnitude 3.3, the minima are about 14.2. Chi Cygni lies near the 4th-magnitude star Eta Cygni. It is markedly red. *See maps* 9, 15.

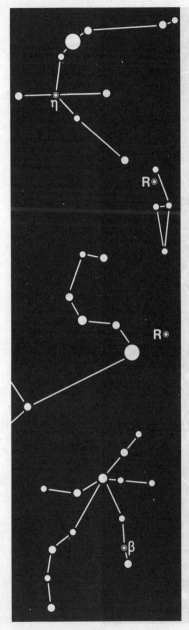

Eta Aquilae is a typical Cepheid with a range from magnitude 3.7 to 4.7 and a period of 7.2 days. It is easy to see as it lies midway between Delta and Theta Aquilae, both of which are just below the 3rd magnitude. *See maps 3, 5, 9, 15.*

R Scuti is the brightest of all the RV Tauri variables; there are alternate deep and shallow minima, which have spells of irregular behaviour. R Scuti ranges from magnitude 5.7 to 8.6, making it a good binocular object. *See maps 3, 5, 15.*

R Leonis is a typical Mira star with a period of 313 days. It is a good binocular object with magnitude ranging from 5.4 to 10.5. With spectral type M it is very red and is visible to the naked eye near maximum. *See maps 1, 11, 13.*

Beta Persei, known as Algol, appears variable but, like Beta Lyrae, is in fact an eclipsing binary. Every 2.87 days the fainter member eclipses the brighter and the magnitude of the pair drops from 2.1 to 3.3. It is at minimum for only 20 minutes. *See maps 2, 6, 11.*

101

Variable stars S. Hemisphere

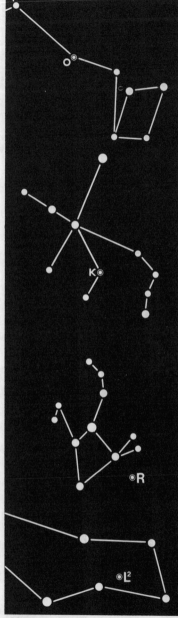

Mira Ceti is the best known and brightest of long-period variables. It lies only 2° south of the celestial equator and is therefore visible from every inhabited country. The period is 331 days, and the range of magnitude is between 1.7 and 10, but the amplitude is not regular. Mira is visible with the naked eye for a few weeks each year. It is strongly red and has a faint white companion. *See maps 5, 9.*

Kappa Pavonis, a short-period variable, has a range from magnitude 4 to 5.5, so that the star is always visible to the naked eye. Period: 9.1 days. Zeta is a good comparison star. *See maps 10, 12, 14, 16.*

R Leporis, a Mira-type, long-period variable, is one of the reddest stars. The period is 432 days. At maximum the magnitude is 5.9 and the minimum the 10.5; not easily visible. *See maps 1, 7, 11, 14.*

L² Puppis, having the declination −44°, is invisible from Europe. It is a red, semi-regular variable. Mean period: 141 days. Magnitude range: 3.4 to 6.2; often clearly visible. *See maps 1, 7, 12, 14.*

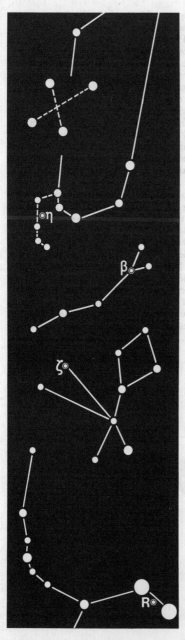

Eta Carinae is a
very erratic variable.
In the 1800s it was
brighter than any star
except Sirius, but is
now below magnitude 7.
For many years it
has been invisible to
the naked eye. In a
telescope it looks like
a red blob associated
with nebulosity. To
northern observers it
never rises above the
European horizon.
The star's nature is
decidedly uncertain.
See maps 10, 12, 14.

Beta Doradûs is a
Cepheid. Its period is
9.8 days; magnitude
range 4.5 to 5.7. The
declination is −62° 30′.
It is too far south to
be seen in Europe or
the United States.
See maps 10, 12, 14.

Zeta Phoenicis, with
declination −55°, is an
eclipsing binary. The
"normal" magnitude is
3.6, the minimum 4.1.
Period: only 1.7 days.
Kappa Phoenicis is a
useful comparison star.
See map 16.

R Centauri lies close
to the two brilliant
pointers to the
Crux Australis and as
such is easy to find,
although for most of the
time it is well below
naked-eye visibility.
See maps 14,16.

103

Clusters

Some of the most spectacular objects in the Galaxy are the star clusters which are of two main types: open and globular. Some, such as the Pleiades or Seven Sisters in Taurus, are visible with the naked eye but most require telescopic aid.

Open or loose clusters have no particular shape and may contain several dozen to several hundred stars. It may be assumed that the stars in a cluster are of about the same age and have a common origin. In some cases the prominent stars are hot and white, with considerable nebular material spread between them, as in the Pleiades, although the interstellar material is not easily visible except by means of photography. Other clusters are much more advanced in their evolution. An open cluster may eventually be so affected by adjacent non-cluster stars that it will disperse and cease to be recognizable.

Globular clusters are different. They are symmetrical systems, containing up to a million stars, surrounding the Galaxy in what is termed the "galactic halo". Out of the hundred known globular clusters only three are clearly visible with the naked eye. Clusters are commonest in the Southern Hemisphere of the sky, giving the first definite proof that the Sun lies well away from the galactic centre. They contain Cepheids and so their distances may be measured. All of them are extremely remote, at least 20,000 light-years away. If the Sun were a member of a globular cluster, the sky would indeed be glorious. There would be many stars bright enough to cast shadows and there would be no real darkness.

Globular cluster
Open cluster

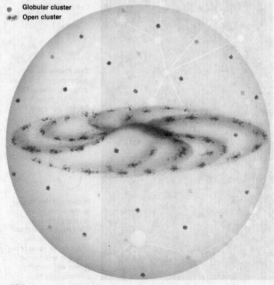

Nebulae

Open clusters N. Hemisphere

Gaseous or galactic nebulae (Latin *nebula*, a cloud) are stellar birthplaces; fresh stars are slowly condensing out of the nebular material. They shine because of the stars in or close to them. If the intermixed stars are extremely hot, they cause the nebular material to emit a certain amount of self-luminosity (emission nebula A); if the stars are cooler, the nebula shines only by reflection (reflection nebula B). If there are no suitable stars available, the nebula cannot shine at all and appears as a dark mass blotting out the light of objects beyond (dark nebula C). But everything depends upon the position of the observer. A nebula that appears dark to one observer may be illuminated on its far side by a star that is hidden from him, so that it would then look bright.

The material in a nebula is highly rarefied. A good example is the famous nebula in Orion's Sword, which is 30 light-years in diameter. If it were possible to take a 2.5 centimetre core sample right through it, the total "weight" of material collected would be less than that of a small coin. Astronomers have also found that many bright nebulae are merely the visible parts of much larger clouds.

The so-called planetary nebulae are neither planets nor nebulae in the true sense but are called planetaries, because they resemble the disks of planets. They are really hot stars surrounded by shells of gas, evolving to the white dwarf stage. Also of interest are the supernova remnants, of which the most famous example is the Crab Nebula. Just as an ordinary nebula is the area in which a star first forms, so a supernova remnant is the result of an ordinary star's death.

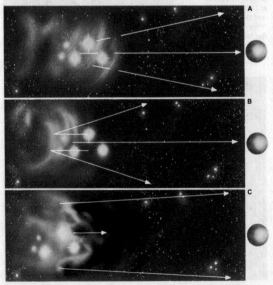

Open clusters N. Hemisphere

There are a number of prominent, naked-eye open clusters in the Northern Hemisphere of the sky. The most striking of these are the two clusters in Taurus: the Hyades, surrounding Aldebaran, and the Pleiades, or Seven Sisters, where the brightest star is the 3rd-magnitude Alcyone. The Hyades are rather overpowered by the brilliant orange light of Aldebaran, which is not a true member of the cluster at all but merely happens to lie about midway between the Hyades and Earth. The Pleiades are striking when seen through binoculars or a low-power telescope. Also of special interest are the two clusters in Cancer: M44 (Praesepe), which is visible with the naked eye and is one of the oldest known clusters, and M67, an easy binocular object. In Perseus there are the twin clusters found in the Sword-handle (not to be confused with Orion's Sword) which, for some reason, were omitted from Messier's list.

M67 (*right*), close to the 6th-magnitude star Alpha Cancri (Acubens), is on the fringe of naked-eye visibility. Praesepe, M44, (*below*) lies close to Delta (magnitude 4) and Gamma Cancri (magnitude 4.7). Unlike the Pleiades, Praesepe has no nebulosity. *See maps 1, 7, 11, 13.*

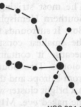

NGC 884 ⁂
NGC 869 ⁂

The twin clusters, Chi and h Persei (*left*), are in the same low-power field and make up one of the most spectacular objects in the sky. *See maps 2, 6, 11.*

⁂ M 45

In the Pleiades cluster, M45 (*below*), six stars are visible to the naked eye. The brightest star is Eta Tauri (Alcyone) with magnitude 2.9. *See maps 2, 7, 11.*

Open clusters S. Hemisphere

The most striking open clusters are to be found in the Southern Hemisphere, the finest example being the Jewel Box. It surrounds the star Kappa Crucis in Crux Australis, the smallest constellation in the sky, and is so named because it contains over one hundred stars of various colours. Two Southern Hemisphere clusters fairly visible from Europe and the United States are M41 and M11. M41 is a bright cluster well worth looking at and just visible with the naked eye. M11, which is known as the Wild Duck, is one of the most beautiful open clusters in the sky; it lies near Lambda Aquilae.

The whole area of Scorpius and Sagittarius is rich in clusters and nebulae and is worth sweeping with binoculars or a low-power telescope, as is the Scutum region which adjoins the "tail" of Aquila.

M7 (*right*) and M6 are two bright clusters near the "sting" of Scorpius. Both can be seen with the naked eye. Neither cluster is clearly visible to observers living north of latitude 40° N.
See maps 3, 10, 14, 16.

NGC 4755, surrounding Kappa Crucis, is one of the most beautiful open clusters in the whole sky. Nicknamed the Jewel Box because it contains stars of varied colours, it is easily seen with binoculars.
See maps 10, 12, 14, 16.

M41, in Canis Major, contains about 50 stars; it is on the fringe of naked-eye visibility. There is a 6th-magnitude star on its south-eastern edge. The cluster is about 1,600 light-years away. *See maps 1, 7, 14.*

M11, nicknamed the Wild Duck in Scutum lies below the "tail" of Aquila to Northern-Hemisphere observers. It is fan shaped with an 8th-magnitude star just inside the apex. The region is very rich. *See maps 3, 5, 9, 15.*

M46, in Puppis, is a spectacular object under good conditions and is rich in faint stars. It is not easy to find in a bright or misty sky. Sufficiently far north, it can be seen in Europe and the USA. *See maps 1, 7, 12, 14.*

Globular clusters N. Hemisphere

Globulars are vast star systems; because they are very remote they are best seen with telescopes of a fairly wide aperture. Globular clusters are much more condensed than open clusters and relatively high powers are needed to resolve them into stars, although smaller instruments are adequate for the outer parts of the brightest globulars. In the Northern Hemisphere, M13 in Hercules is the only globular sometimes visible to the naked eye and is best seen against a dark, transparent sky; binoculars also show it clearly. All the other globular clusters shown here are also visible with binoculars, although they appear only as misty patches. Most of the globular clusters lie in the Southern Hemisphere of the sky, including the two brightest of all, Omega Centauri and 47 Tucanae, because of the off-centre position of the Sun in the Galaxy.

Of the two globulars in Hercules, M13 (*above*) lies between Eta and Zeta; with magnitude 5.7 it is a naked-eye object. M92 (*right*) is less easy to find. The magnitude is 6.1. *See maps 2, 3, 6, 9, 15.*

110

☀M5

M5 (*left*), in Serpens, has an integrated magnitude of 6.2, making it a fine telescopic object. Its diameter is 130 light-years. More than 100 RR Lyrae variables have been discovered. *See maps 3, 5, 15.*

M 15 ☀

M15, in Pegasus, (*above*), is close to the 2nd-magnitude star Epsilon Pegasi. The centre is bright and the outer parts are occasionally seen. It is a fine globular, comparable to M13. *See maps 4, 5, 8, 9.*

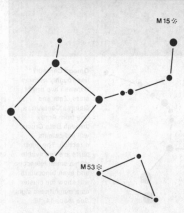

M 53 ☀

M53, in Coma Berenices, is within one degree of the 4th-magnitude star Alpha Comae. It has a bright nucleus and is very regular, making it a splendid object. The integrated magnitude is 7.6. *See maps 4, 8, 13.*

111

Globular clusters S. Hemisphere

The Southern Hemisphere is comparatively rich in globular clusters, particularly in the regions of Scorpius and Sagittarius. Easily visible clusters include M4, M80 and M22. Two other globulars in the Southern Hemisphere, Omega Centauri and 47 Tucanae, far surpass the finest globular in the Northern Hemisphere—that of M13 in Hercules. Omega Centauri takes pride of place being the brightest. It is prominently visible as a misty patch and is a glorious sight through even a small telescope. It lies at a distance of 22,000 light-years and contains thousands of 12th- to 15th-magnitude stars. Almost as spectacular as Omega Centauri is 47 Tucanae which borders the Small Magellanic Cloud. It is of the 5th magnitude and is easily visible to the naked eye. It contains stars of magnitude 12 to 14 and below.

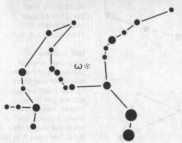

Omega Centauri lies roughly midway between two bright stars, Zeta and Gamma Centauri; a line from Acrux through Beta Crucis will indicate its direction. The outer parts are resolvable with a small telescope and even binoculars will show the cluster as a magnificent sight. *See maps 14, 16.*

NGC 104

NGC 104, 47 Tucanae, like Omega Centauri, is easily visible to the naked eye. It is not far from the Small Cloud of Magellan. *See maps* 10, 12, 14.

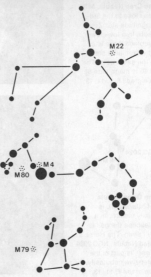

M22 (*above*), a bright globular, is close to the 3rd-magnitude star Lambda Sagittarii. Its declination is −24° so it never rises to a convenient altitude over northern Europe or the northern USA. It is an easy binocular object and is not hard to resolve to its centre. *See maps* 3, 5, 10.

M22

M4 (*right*) is easy to locate because it is near the supergiant Antares. M80, a splendid globular, lies between Antares and Beta Scorpii. *See maps* 3, 10, 14, 16.

M80   M4

M79 (*right*), in Lepus, is at declination −24°. It is rather small with a bright centre. The nearest bright star is Beta Leporis. *See maps* 1, 7, 14.

M79

Nebulae N. Hemisphere

Bright nebulae are of various types. In recent years it has been found that the visible objects may be no more than parts of much larger "clouds" which cannot be seen directly. Such is the Great Nebula in Orion, M42, the most famous example of a galactic nebula. It has a diameter of 25 to 30 light-years and is the visible part of an extensive nebulosity covering much of Orion. M42 itself lies about five degrees south of the celestial equator. Also in this area is the Horse's Head, a dark nebula that is not easy to see with a small telescope, but it is spectacular when photographed with large instruments. Other dark rifts may be seen in the Milky Way in the region of Cygnus.

The Crab Nebula in Taurus, M1, is of an entirely different type and is the best example of a supernova remnant. It is not hard to find but is a dim object in small telescopes. The Northern Hemisphere also includes some fine examples of planetary nebulae which are inappropriately named, because they are certainly not planets and are not true nebulae. The Ring, M57 in Lyra, is particularly easy to locate between the bright stars Beta and Gamma; M27, in Vulpecula, is also distinct, within a few degrees of Gamma Sagittae.

The Crab Nebula, M1, lies close to the 3rd-magnitude star Zeta Tauri. In a low-power eyepiece there is a 6th-magnitude star in the same field. The Crab is magnitude 8½. *See maps 6, 7, 11, 13.*

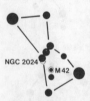

The Orion Nebula, M42 (*far right*), is easily visible; in a telescope the gas is well seen. The Horse's Head Nebula, NGC 2024 (*right*), is part of the extensive nebulosity. *See maps 7, 11, 13.*

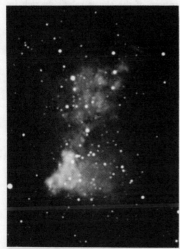

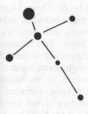

The Dumb-bell Nebula, M27, is in the small constellation of Vulpecula. An unusual planetary nebula, it is a fine object. With a 15 cm reflector the dumb-bell form is well seen. The central star is of magnitude 12. *See maps 3, 5, 9, 15.*

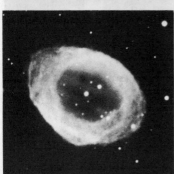

The Ring Nebula in Lyra, M57, a more typical planetary nebula, lies midway between Beta and Gamma Lyrae. The central star is faint and a telescope of at least 25 cm aperture is needed. *See maps 2, 6, 9, 15.*

Nebulae S. Hemisphere

The Southern Hemisphere of the sky is relatively rich in bright nebulae, particularly in the constellations of Sagittarius and Scorpio, lying in the direction of the galactic centre. Two objects are of interest. The nebula surrounding the erratic variable Eta Carinae is extremely complex and a magnificent sight. Whether or not Eta Carinae itself will return to its former eminence, when it outshone even Canopus, remains to be seen (since the late nineteenth century it has remained below naked-eye visibility). The Looped Nebula, sometimes described as the Tarantula Nebula, around 30 Doradûs, is part of the Large Cloud of Magellan and is about 160,000 light-years away. If it were as close to Earth as the Great Nebula in Orion it would be brilliant enough to cast perceptible shadows.

NGC 2070

The Looped Nebula, NGC 2070, in the Large Cloud of Magellan, is easily seen with binoculars or low magnification on a telescope. *See maps* 10, 12, 14, 16.

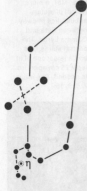

η

The Keyhole Nebula around Eta Carinae is one of the showpieces of the southern sky. The whiteness of the nebulosity contrasts with the orange-red of Eta Carinae itself. *See maps* 12, 14.

Three bright galactic nebulae.
The Trifid Nebula, M20 (*top*), in
Sagittarius is a typical emission
nebula, more or less symmetrical.
The Lagoon Nebula, M8 (*centre*),
also in Sagittarius is an emission
nebula associated with a cluster.
The Omega or Horseshoe Nebula,
M17 (*bottom*) is now recognized to
be part of a much larger cloud
that cannot be seen directly.
See maps 3, 5.

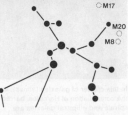

117

Galaxies

Our Galaxy is a typical spiral and is certainly not unique. Many other spirals are known and formerly they were incorrectly termed spiral nebulae. But not all galaxies take this form. Some are irregular and two of the brightest of these are the southern Clouds of Magellan.

The Local Group, of which Earth's Galaxy is a member, contains a few large systems and more than twenty smaller ones. The Andromeda Spiral, at a distance of 2.2 million light-years, is a member; so are the Clouds of Magellan, a spiral in Triangulum, and another massive system, Maffei 1, which is hard to see because it is so heavily obscured by material in the plane of Earth's Galaxy. Beyond the Local Group the distances become immense and astronomers know of galaxies that are more than 5,000 million light-years away. Distances of the nearer galaxies can be measured by the Cepheid method. With more remote systems the Cepheids fade into the background blur and the measurement of distances becomes much more difficult.

Apart from the normal spirals, there are "barred spirals" in which the arms come from the ends of a bar through the main plane. There are also elliptical and irregular systems. Some galaxies emit strong ultraviolet or radio

In this cluster of galaxies (*above*), in the constellation of Hercules, barred spirals and elliptical galaxies are seen. The clustering is genuine, and the individual galaxies are not receding from each other.

waves; for example, the Seyfert systems have condensed active nuclei and inconspicuous spiral arms.

Galaxies tend to occur in clusters (not to be confused with star clusters), some of which are much more populous than Earth's Local Group. The Virgo cluster, for example, at about 65 million light-years, contains hundreds of systems. With remote groups even a massive galaxy will appear as nothing more than a tiny smudge of light. Then there are the extraordinary objects known as quasars, which are extremely remote and super-luminous. Whether or not these are directly associated with normal galaxies is not yet known.

Apart from the members of the Local Group, all the galaxies are racing away from Earth, so that the entire universe is expanding. The most remote objects known are moving away at more than 90 per cent of the velocity of light.

In 1781 the French astronomer C. Messier drew up a list of more than 100 clusters, nebulae and galaxies and the "M" numbers are still used; the Andromeda Spiral, for example, is M31. Other systems of nomenclature have since been introduced but most of the brightest objects were included in Messier's list.

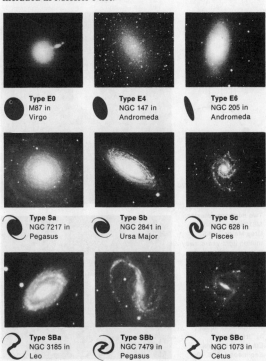

Type E0
M87 in
Virgo

Type E4
NGC 147 in
Andromeda

Type E6
NGC 205 in
Andromeda

Type Sa
NGC 7217 in
Pegasus

Type Sb
NGC 2841 in
Ursa Major

Type Sc
NGC 628 in
Pisces

Type SBa
NGC 3185 in
Leo

Type SBb
NGC 7479 in
Pegasus

Type SBc
NGC 1073 in
Cetus

Galaxies N. Hemisphere

External galaxies, far beyond our own system, are so remote that only a few can readily be seen in the Earth's skies. Many galaxies are difficult to locate because they lie in the direction of the galactic plane and are obscured by interstellar material, for example, Maffei 1 and Maffei 2, which were discovered in 1968, when they were named in honour of their discoverer, the Italian astronomer Paolo Maffei. Astronomers now believe that Maffei 1, which appears to be a giant elliptical system, is a member of the Local Group. It lies at a distance of three million light-years in the constellation of Cassiopeia. Maffei 2 seems to be a spiral of lower mass which lies well beyond the Local Group at a distance of 15 million light-years.

There are, however, some northern galaxies that are well away from the plane of our own system. M31, the Great Spiral in Andromeda, is an example. Unfortunately it is placed at an angle and its full beauty is lost. In contrast, the Whirlpool Galaxy, M51 in Canes Venatici, is face-on. It is more than 35 million light-years away. If it were as close as the Andromeda Spiral it would be truly imposing. It is easy to locate although telescopes of considerable aperture are needed to show the spiral form.

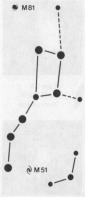

Two northern galaxies. M81 (*above right*), in Ursa Major, is 8.5 million light-years away. It is not a difficult object because its integrated magnitude is above 8. Near by is M82, an irregular, active galaxy and a strong radio source. M51 (*right*), the Whirlpool, is about $3\frac{1}{2}°$ from Alkaid, Eta Ursae Majoris. *See maps 4, 6, 8, 13.*

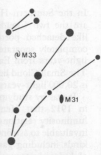

The two great northern spirals. M33 (*left*), in Triangulum, is not easy to detect, even with binoculars or a small telescope. M31 (*below*), the Great Spiral in Andromeda, appears as an elongated blur of light and can easily be found with binoculars. *See maps 2, 4, 8, 9.*

Galaxies S. Hemisphere

In the Southern Hemisphere the most interesting objects are the two Clouds of Magellan. Superficially they look like detached portions of the Milky Way but they are completely separate systems. The Large Cloud is 160,000 light-years from Earth and 40,000 light-years in diameter. The Small Cloud lies at a distance of 190,000 light-years and is 20,000 light-years in diameter.

It was by studying Cepheid variables in the Small Cloud in 1912 that Henrietta Leavitt discovered the period-luminosity relationship for these stars which has proved so invaluable to astronomers. The Clouds contain objects of all kinds including gaseous and planetary nebulae, open and globular clusters and extremely brilliant stars. S Doradûs, in the Large Cloud, is a giant star in one of the final stages of evolution. Although it is at least one million times more luminous than the Sun its great distance prevents it from being visible without a telescope.

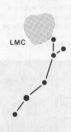

The Large Cloud of Magellan is a naked-eye object lying between Dorado (*above*) and Hydrus. It is classed as a dwarf irregular galaxy, although there are vague signs of a bar.
See maps 10, 12, 14, 16.

The Small Cloud of Magellan, or Nubecula Minor, is of the same type as the Large Magellanic Cloud, although it is a separate system. It lies in Tucana. *See maps 10, 12, 14.*

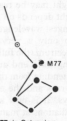

M77, in Cetus, is a Seyfert galaxy—a galaxy with a bright, condensed nucleus and faint arms. The total magnitude is about 9. With a moderate-sized telescope it is easy to find. *See maps 5, 11.*

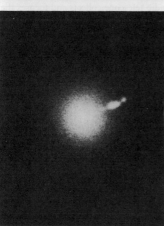

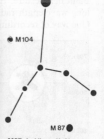

M87, in Virgo (*left*), is an elliptical galaxy, extremely massive and a strong source of radio waves. M104 (*below*) is an S6-type spiral, nicknamed the "Sombrero Hat". *See maps 1, 13.*

Invisible astronomy

Light may be regarded as a wave motion and the colour of light depends on its wavelength; for visible light, red has the longest wavelength and violet the shortest. Visible light takes up only a very small part of the total electromagnetic spectrum (the full range of wavelengths). Extending from the long-wave end are the infra-red and then the radio waves; extending from the short-wave end there are ultra-violet rays, X-rays and the extremely short, highly penetrative gamma rays. It is now known that various bodies in the sky emit radiations at many different wavelengths and therefore what may be termed "invisible astronomy" has become of paramount importance.

Radio waves from the Milky Way were first detected in 1931 by Karl Jansky in the United States, admittedly by accident: he was using a home-made aerial to investigate the causes of static. Subsequently, radio telescopes were built, of which the most famous is the 76.2-metre "dish" at Jodrell Bank in Cheshire. A radio telescope collects and focuses the long-wavelength radiations but produces no visible picture. One way of recording is by a pen on a moving strip of paper.

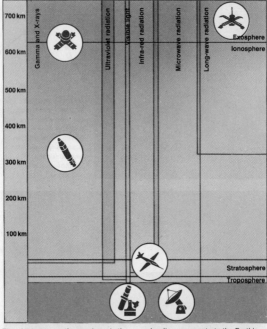

The electromagnetic spectrum is the complete range of electromagnetic radiations from the shortest (gamma rays) to the longest (radio) waves. Only two bands of wavelengths, light and radio, can penetrate the Earth's atmosphere and therefore be analysed from the ground. Balloons, aircraft and satellites are used to study all other wavelengths.

The Sun is a radio source and so is Jupiter but most other radio sources are much more remote. Some, such as the Crab Nebula, are supernova remnants and others lie well beyond the Milky Way. In particular, radio astronomy has added considerably to the knowledge of more distant objects in the universe, such as the extra-galactic quasars, possibly the most powerful objects known to mankind.

Infra-red astronomy has also become highly important, particularly in studies of very young stars still surrounded by dusty "cocoons". Short-wave astronomy is hampered by the effects of the Earth's atmosphere, which acts as an effective screen (this also affects some radio wavelengths). X-rays from the sky were first detected in 1962 by means of equipment carried in a rocket. Subsequently, artificial satellites were launched specifically to study X-ray sources and the results have been most interesting. The Crab Nebula, for example, is an X-ray source. Gamma rays from the sky were also identified by rocket equipment in 1969, but gamma-ray astronomy is still in its infancy. The most powerful gamma-ray source discovered is the pulsar in Vela.

The active galaxy M82, in Ursa Major (*above*), is a powerful radio source. Lying 10.5 million light-years away, there are indications that a tremendous explosion took place inside the nucleus of the system 1.5 million years ago. Analyses have revealed intricate and immense hydrogen structures moving at up to 160 km per second.

The Jodrell Bank 76.2 m (250 ft) dish (*above*) is one of the most famous radio telescopes in the world. Used in stellar research, it is one of the largest to be fully steerable.

The British-built satellite UK6 (Ariel 6) was launched in 1979. It was specifically designed to study X-ray sources such as unusual binary systems and some galaxies.

Cosmology

How did the universe begin? How is it evolving and how will it die—even assuming that it will come to an end? These are questions that cannot yet be answered but at least astronomers have learned more in the last decade than would have seemed possible.

The first great breakthrough was made in 1923 when Edwin Hubble, at the Mount Wilson Observatory, proved that the objects formerly termed spiral nebulae were independent galaxies. Hubble detected Cepheid variables in some of the systems. He was able to obtain the distances of the Cepheids by observing their periods and by using the period-luminosity law, and it was at once clear that our Galaxy was only one of many. Hubble also confirmed that, apart from the members of the Local Group, all the galaxies appear to be receding from each other and that the farther away they are, the faster they go.

The essential clue came from what is termed the Doppler Effect. When a light-source approaches Earth, the wavelength of the light appears to be shortened and the light will seem to be "too blue". When the light-source is receding, the wavelength is lengthened, and the light appears to be "too red". This effect shows up in the spectrum of a galaxy (or, for that matter, of a star or planet). If the dark lines in the spectrum are moved over towards the red end of the rainbow band, then the object must be receding and the amount of red shift is a key to the velocity of recession. Astronomers are now aware of galaxies that are moving away at appreciable fractions of the velocity of light. There are also the enigmatical quasars, which may well be even more remote and fast-moving. One quasar, known by its catalogue number of OQ 172, has a recessional velocity estimated to be 91 per cent that of light. If the law of higher velocity with greater distance holds good indefinitely, then a point must be reached at which a galaxy is moving away at the velocity of light. This means that it cannot be observed and that there will be a limit to the boundary of the observable universe, although not necessarily to that of the universe itself. Every group of galaxies is receding from every other group and the recession of galaxies is interpreted as a general expansion of the universe and leads to three main theories on the evolution of the universe.

Calculating "backwards", it seems that the expansion began more than 10,000 million years ago. Perhaps 15,000 million years is a better estimate—although there is no proof that the rate of expansion then was the same as it is now, so that uncertainties are bound to remain.

When astronomers begin to consider the origin of the universe certain difficulties immediately arise. According to one theory, everything was created at one moment in a kind of "primeval atom"; expansion began and galaxies were formed, out of which stars were born. This is popularly known as the Big Bang theory although it cannot hope to explain how the material was created in the first place. Another explanation, the so-called Steady State theory, was popular for some years. It assumed that the universe has always existed and will exist for ever so that, as old galaxies die or move beyond the range of the observable universe, they are replaced by new ones, produced from matter that appears spontaneously out of nothingness. The universe would therefore be in a steady state. So many objections have been advanced to this theory that it has now been abandoned by almost all astronomers.

The third possibility is that the present phase of expansion will be followed by a period of contraction. The galaxies will come together again and there will be another Big Bang. This is the Cyclic or Oscillating Universe theory. If it is correct there should be a Big Bang every 80,000 million years or so. Everything depends upon whether the density of material in the universe as a whole is great enough for its gravitational effect to prevent the expansion from continuing indefinitely. The critical value is about three hydrogen atoms in every cubic metre. If the mean density is lower than this, then the expansion cannot be stopped. Present research indicates that the density is below this critical point.

One important clue to the understanding of cosmology has been the discovery of "background radiation" coming from all directions, possibly the remnant of the original Big Bang. Astronomers at present have to admit that they do not know how the universe began. Whether they will ever find out remains to be seen.

Time chart of astronomy

1543	Nicolas Copernicus' book published, in which he claimed that the Earth moves round the Sun.
1609	First serious telescopic observations by Galileo Galilei.
1618	Kepler's laws of planetary motion published.
1668	First reflector made, by Sir Isaac Newton (this is the probable date; it was certainly in existence by 1671).
1675	Royal Greenwich Observatory founded.
1687	Sir Isaac Newton's *Principia* published.
1781	Uranus discovered, by Sir William Herschel. Charles Messier's catalogue of clusters and nebulae published.
1786	First approximately accurate idea of the shape of the Galaxy discovered, by Sir William Herschel.
1801	First asteroid discovered: Ceres, by G. Piazzi at Palermo.
1802	Existence of binary star systems established, by Sir William Herschel.
1814	Dark lines in the solar spectrum studied, by J. Fraunhofer.
1834/8	First extensive observations of the stars in the far Southern Hemisphere made, by Sir John Herschel.
1838	First measurement of distance of a star, by F. W. Bessel.
1845	Spiral nature of the galaxies discovered, by Lord Rosse with his 182 cm reflector at Birr Castle in Ireland.
1846	Neptune discovered.
1859	Dark lines in the solar spectrum explained, by G. Kirchhoff.
1863/4	First classifications of the spectra of stars, by A. Secchi in Italy and Sir W. Huggins in England.
1912	Period-luminosity law of the Cepheid variables discovered, by Miss H. Leavitt at Harvard.
1915	Albert Einstein's general theory of relativity published.
1917	The 254 cm reflector at Mount Wilson completed.
1923	Proof given, by E. Hubble, that the galaxies are independent systems rather than parts of our Galaxy.
1929	Velocity/distance relationship of galaxies announced, by E. Hubble.
1930	Pluto discovered, by Clyde Tombaugh.
1931	First experiments on receiving radio waves from the sky, by K. Jansky in the USA.
1938	New (and correct) theory of stellar energy proposed independently, by H. Bethe and G. Gamow.
1942	Radio waves from the Sun detected, by J. S. Hey and his colleagues.
1946	The radio source Cygnus A identified.
1948	The 508 cm reflector at Palomar in California completed.
1952	Error in the Cepheid scale discovered, by W. Baade.
1955	The 7,620 cm radio "dish" at Jodrell Bank completed.
1963	Quasars identified, by M. Schmidt at Palomar.
1965/6	The 3° K microwave radiation detected, by A. Penzias and R. Wilson.
1967	First pulsar identified, by Miss J. Burnell at Cambridge.
1969	First optical identification of a pulsar, in the Crab Nebula.
1974	The 389 cm reflector at Siding Spring, Australia, completed.
1976	The 600 cm reflector, USSR, completed.
1977	Rings of Uranus discovered. Discovery of Chiron. Discovery of Charon (satellite of Pluto).
1983	Full survey of the sky in infra-red carried out by IRAS. Discovery of "infra-red excesses" with certain stars, possibly indicating planet-forming material.
1985/6	Return of Halley's Comet.

Time chart of space research

1895 First important papers about space flight published, by K. E. Tsiolkovskii.

1926 First liquid-propelled rocket launched, by R. H. Goddard.

1949 First step-rocket fired, from White Sands in New Mexico.

1957 4 October. First artificial satellite launched: Sputnik 1 (USSR).

1958 First successful American artificial satellite: Explorer 1. Its instruments detected the Earth's Van Allen Belts.

1959 First successful lunar probes (USSR).

1961 First manned space flight: Yuri Gagarin in Vostok 1 (USSR).

1962 First American to orbit the Earth: John Glenn.
First successful planetary probe: Mariner 2, to Venus (USA).

1964 First good close-range pictures of the Moon, obtained from the American probe, Ranger 7.

1965 First successful Mars probe: Mariner 4 (USA). Close-range pictures obtained, altering all existing ideas about Mars.

1966 First soft landing on the Moon by an automatic probe: Luna 9 (USSR).
The first Lunar Orbiter (USA), sent back high-quality, close-range pictures of the Moon's surface.

1967 First soft landing of an unmanned probe on Venus: Venera 7 (USSR).

1968 First manned flight around the Moon: F. Borman, J. Lovell and W. Anders in Apollo 8 (USA).

1969 21 July. First manned landing on the Moon: N. Armstrong and E. Aldrin in Apollo 11 (USA).
Improved pictures of Mars sent from Mariners 6 and 7 (USA).

1971 First capsule landed on Mars, from the Soviet probe Mars 2; loss of contact prevented any useful results.
Mariner 9 put into orbit around Mars, and, until mid 1972, sent back thousands of pictures; the great volcanoes were seen for the first time.

1972 Apollo programme ended (Apollo 17).

1973/4 Operational "lifetime" of the American space station Skylab, manned by three successive crews. Much astronomical work carried out.

1973 First successful probe to Jupiter: Pioneer 10 (USA) bypassed the planet and sent back information from close range.

1974 First pictures of Venus from close range: Mariner 10 (USA). First successful probe to Mercury: also Mariner 10, which made three active passes of the planet, two in 1974-5. Second successful Jupiter probe: Pioneer 11.

1975 First pictures received from surface of Venus: automatic probes Veneras 9 and 10 (USSR).

1976 First successful soft landings on Mars: Vikings 1 and 2 (USA). Much information was obtained.

1978 Several probes sent to Venus: Veneras 11 and 12 (USSR) and two Pioneers (USA). Further information about the planet obtained, but no more pictures from the surface.

1979 Voyager 1 bypassed Jupiter; active volcanoes discovered on Io.

1980 Voyager 1 passed Saturn.

1981 First flight of the NASA Space Shuttle.

1985 Launch of probes to Halley's Comet.

Glossary 1

Albedo The reflecting power of a non-luminous body. A perfect reflector would have an albedo of 100 per cent.

Ångström unit The hundred-millionth part of a centimetre.

Antoniadi scale A roman numeral indicates the quality of the seeing according to the following scale:

I Perfect seeing, without a quiver

II Slight undulations, with moments of calm lasting several seconds

III Moderate seeing, with larger air tremors

IV Poor seeing, with constant troublesome undulations

V Very bad seeing, scarcely allowing the making of a rough sketch

Aperture The diameter of an opening through which light passes in an optical instrument.

Aphelion The furthest distance of a planet from the Sun.

Apparent magnitude The apparent brightness of a celestial object: the lower the magnitude, the brighter the object.

Aurora (Polar Lights) A diffuse glow in the upper air caused by electrified particles emitted from the Sun.

Axis An imaginary line about which a body rotates. The polar diameter of a planet is the axis of rotation.

Binary Two stars that move around their common centre of gravity.

Black hole The remains of a massive star after its final collapse and contraction into a state where the gravitational pull is so strong that not even light can escape.

Caldera A volcanic crater.

Celestial sphere An imaginary sphere surrounding the Earth, concentric with the Earth's centre.

Cepheid variable A variable star of short period. The fluctuations are regular and are linked with its real luminosity; the longer the period, the more luminous the star.

Chromosphere That part of the Sun's atmosphere that lies just above its visible surface or photosphere.

Circumpolar star A star that never sets. Ursa Major, for example, is circumpolar over the British Isles and Crux Australis is circumpolar over New Zealand.

Conjunction (1) A planet is said to be in conjunction with a star when it is apparently close to it in the sky. (2) The planets within the Earth's orbit, Mercury and Venus, are at inferior conjunction when lined up between Earth and Sun and at superior conjunction when on the far side of the Sun. It follows that Mars and the other planets outside the Earth's orbit can only reach superior conjunction.

Constellation A group of stars within an imaginary outline.

Corona The outermost part of the Sun's atmosphere. Made up of very thin gas, it is invisible to the naked eye except during a total eclipse.

Cosmology The study of the universe.

Culmination The maximum altitude of a celestial body.

Declination The angular distance of a celestial body from the celestial equator.

Density The quantity of matter contained within a unit of volume.

Direct motion The movement of a celestial body from west to east, that is, in the same direction as that of the Earth, around the Sun.

Doppler Effect The apparent change in the wavelength of light according to the motion of the body emitting it, in relation to the observer's position. With an approaching light source the wavelength is shortened ("too blue"), with a receding source it is lengthened ("too red").

Double star A pair of stars. A double may be caused by a genuine physical association (when it is known as a binary star) or by an optical trick: two stars appearing to be close together but in fact they just happen to lie in almost exactly the same line when seen from Earth.

Eccentricity The measure of how well a planet's orbit compares to a perfect circle.

Eclipse, lunar The passage of the Moon through the shadow cast by the Earth. Lunar eclipses may be either total or partial.

Eclipse, solar The covering of the Sun by the Moon, when seen from Earth. Solar eclipses may be either total, partial or annular. An annular eclipse occurs when the Moon is close to its point of maximum recession from the Earth and so is too small to hide the Sun completely.

Eclipsing binary A binary star, one component of which is seen to pass in front of the other, thereby cutting out some or all of its light.

Ecliptic The apparent yearly path of the Sun against the stars.

Elongation The angular distance of a planet from the Sun or of a satellite from its primary planet.

Equator, celestial The projection of the Earth's equator on to the celestial sphere, thus dividing the sky into equal hemispheres.

Equinox The two points at which the Sun crosses the celestial equator; the spring equinox (first point of Aries) is reached about 21 March and the autumnal equinox about 22 September.

Escape velocity The minimum velocity that an object must possess to escape from the surface of a planet or other body.

Extinction The apparent reduction in the brightness of a star or planet when low over the horizon because more of its light is absorbed by the Earth's atmosphere.

Eyepiece The lens (or lenses) at the eye end of a telescope responsible for enlarging the image produced by the object-glass (for a refractor) or mirror (for a reflector).

Faculae The bright patches on the Sun's photosphere.

Finder A small, wide-field telescope attached to a larger one for locating objects in the sky.

Flare, solar Brilliant outbreaks in the solar atmosphere normally detectable only by spectroscopic methods.

Flare star A faint red star that has short-lived explosions on its surface. These explosions cause the star to appear temporarily brighter.

Fraunhofer lines The dark lines in the spectrum of the Sun.

Galaxy A star system. Most galaxies are so remote that their light takes millions of years to reach Earth.

Inclination Measure of the tilt of a planet's orbital plane, in relation to that of the Earth's.

Inferior planet Any planet that orbits between the Sun and the Earth; that is, Mercury and Venus.

Libration An effect caused by the apparent slight "wobbling" of the Moon from side to side, as seen from Earth. As a result a total of 59 per cent of the Moon's surface can be observed although no more than 50 per cent at any one time. The remaining 41 per cent remained unknown to observers from Earth until the Space Age explorations.

Light-year The distance travelled by light in one year: that is, 9.46 million million kilometres.

Limb An edge or border, as of the Sun, Moon or any planet.

Local Group The group of which our Galaxy is a member. There are more than two dozen systems, including the Andromeda Spiral and the two Clouds of Magellan.

Luminosity The amount of light emitted from a star.

131

Glossary 2

Lunation The interval between one new Moon and the next: that is, 29 days 12 hours 44 minutes.

Magnetosphere The area around a planet in which its magnetic field is dominant.

Magnitude Brightness, according to a scale in which the most brilliant stars are of the first magnitude.

Meridian An imaginary circle through the north and south poles of the celestial sphere.

Meteor A small particle that burns away in the Earth's upper air. It is often known as a shooting star.

Meteorite A natural body, probably associated with asteroids, that is able to reach ground level without being destroyed.

Nebula A cloud of dust and gas in space, from which fresh stars are created.

Neutron A fundamental particle with no electrical charge.

Neutron star A star made up principally, or completely, of neutrons. Theoretically, it is the remnant of a massive star that has exploded. Neutron stars that send out rapidly varying radio waves are known as pulsars.

Nova A star that suddenly flares up to many times its normal brightness and remains brilliant for a limited period before fading back to obscurity.

Object-glass (or Objective) The main lens in a refracting telescope.

Oblateness The degree of flattening at the poles of a celestial body.

Occultation The concealment of one celestial body by another. Strictly speaking, a solar eclipse is an occultation of the Sun by the Moon.

Opposition The position of a planet when exactly opposite the Sun in the sky, as seen from Earth. The planet is then best placed for observation.

Orbit The path of a celestial body.

Parallax The apparent shift in position of an object when viewed from two different positions.

Perihelion The closest point that a planet (or other body) comes to the Sun.

Periodic time *see* Sidereal period.

Phase The apparent change in the shape of the Moon and inferior planets, according to the amount of sunlit hemisphere turned towards the Earth. New Moon, for example, is when the unlit side of the Moon is visible. Full Moon is when its surface is fully exposed as viewed from Earth.

Photosphere The brilliant visible surface of the Sun.

Planetary nebula A small, hot star surrounded by a shell of gas.

Poles, celestial The north and south points of the celestial sphere.

Precession The apparent slow movement of the celestial poles caused by a real shift in the direction of the Earth's axis.

Prominence A mass of glowing gas, chiefly hydrogen, above the Sun's photosphere.

Pulsar *see* Neutron star.

Quasar A very remote, super-luminous object. The nature of a quasar is still uncertain.

Radial velocity The movement of a celestial body towards or away from the observer.

Radiant The point in the sky from which a meteor shower appears to emanate.

Red giant The stage in the evolution of an ordinary star when the core contracts, the surface expands to about 50 solar radii and its temperature drops, giving the star its red colour.

Retrograde motion The movement of a celestial body from east to west; that is, in the opposite direction to that of the Earth.

Right ascension The time that elapses between the culmination of the first point of Aries and the culmination of a celestial body.

Seyfert galaxy A galaxy that has a small, bright nucleus and faint, spiral arms. It is often a strong radio source.

Sidereal period The revolution period of a planet around the Sun or that of a satellite around its primary. Also known as Periodic time.

Solar wind Charged particles from the Sun that travel into the solar system at about 1½ million kph.

Spectroscopic binary A very close double that is only recognizable by its displacing effects in the combined spectrum of the two stars.

Spectrum The range of colour produced by a prism.

Supergiant The stage in the evolution of a massive star when the core contracts, the surface expands to about 500 solar radii and its temperature drops, giving the star its red colour.

Superior planet Any planet beyond the orbit of the Earth in the Solar System.

Supernova A massive star that reaches a peak of luminosity and then explodes in a cataclasmic outburst and dies, leaving a neutron star surrounded by a cloud of expanding gas.

Synodic period The interval between successive oppositions of a superior planet.

Terminator The boundary between the illuminated and dark portions of a planet or satellite.

Transit The passage of a celestial body across the observer's meridian; the apparent passage of a smaller body across the disk of a larger one.

Van Allen Belts Radiation zones of charged particles surrounding the Earth.

Variable star A star that fluctuates in brilliancy. Eclipsing binaries are categorized under this heading.

White dwarf A small, very dense star that has used up its nuclear energy and collapsed. White dwarfs have been described as "bankrupt stars".

Yellow dwarf An ordinary star such as the Sun at a comparatively stable and long-lived stage in its evolution.

Zenith The observer's overhead point (latitude 90°).

Zodiac A belt stretching 8° to either side of the ecliptic, in which the Sun, Moon and all planets are always to be found.

Zodiacal light A cone of light rising from the horizon after sunset or before sunrise. It is caused by sunlight on thinly spread interplanetary material in the main plane of the Solar System.

KEY TO ABBREVIATIONS AND SYMBOLS

Aqr.	Aquarius	Lib.	Libra	Psc.	Pisces
Ari.	Aries	Lyr.	Lyra	s.	second(s)
a.u.	astronomical	M	Messier number	Sco.	Scorpius
	units	m.	minute(s)	Sex.	Sextans
Boo.	Boötes	NGC	New General	Sgr.	Sagittarius
Cap.	Capricornus		Catalogue	Tau.	Taurus
Cet.	Cetus	Oph.	Ophiuchus	TLP	Transient Lunar
Cnc.	Cancer	Ori.	Orion		Phenomena
CrA.	Corona Austrinus	Per.	Perseus	UMi.	Ursa Minor
Cyg.	Cygnus	Phe.	Phoenix	UT	Universal Time
Gem.	Gemini	PsA.	Piscis Austrinus	Vir.	Virgo

//	second(s)	**Types of galaxies**
⁛	nebulae	● E0 🖝 Sa ⌇ SBa
⁛⁛	open/globular clusters	◣ E4 ◆ Sb ℮ SBb
⊙	variable stars	◖ E6 ◗ Sc ⌇ SBc

133

Stellar types 1

DOUBLE STARS NORTHERN HEMISPHERE

Star	Magnitudes	Separation"	Comments
γ Andromedae	2.3, 6.1	10.0	Orange and bluish. Easy
γ Arietis	4.8, 4.7	8.2	Lovely, easy pair. Optical
θ Aurigae	2.7, 7.5	3.0	Test for small telescopes
μ Boötis	4.5, 6.7	109.0	Very wide and easy
ζ Cancri	5.1, 6.0	5.9	Easy. Each star is again a close double
α Canum Venaticorum	2.9, 5.4	19.7	Very easy
η Cassiopeiae	3.6, 7.5	10.1	Very easy. Slow binary
ι Cassiopeiae	4.7, 7.0, 7.1	2.3, 8.2	Triple star
δ Cephei	var, 7.5	41.0	Famous variable. Optical
β Cygni	3.2, 5.4	34.6	Yellow, blue. Loveliest of all doubles. Very wide and easy
γ Delphini	4.5, 5.5	10.4	Easy, contrasting colours
ν Draconis	4.9, 5.0	62.0	Separable with binoculars
α Geminorum	2.0, 3.0	2.2	Castor. Binary, 350 years. Now closing, fairly easy
α Herculis	var, 5.4	4.6	Not difficult
ζ Herculis	3.1, 5.6	below 2	Binary, 34 years. Close
γ Leonis	2.6, 3.8	2.6	Yellowish, bluish. Fine pair
β Lyrae	var, 7.8	46.6	Optical. The famous variable
ε Lyrae	5.1, 6.0, 5.1, 5.4	2.8, 2.3	Quadruple star, two pairs. Two components visible to the naked eye
ζ Lyrae	4.3, 5.9	43.7	Very wide and easy
ζ Persei	2.9, 9.4	13.0	Easy
α Piscium	4.3, 5.2	2.1	Slow binary, not too easy
θ Serpentis	4.5, 4.5	22.6	Very easy
ζ Ursae Majoris	2.4, 3.9	14.5	Naked-eye pair with Alcor
α Ursae Minoris	2.0, 9.0	18.3	Test for small telescope

DOUBLE STARS SOUTHERN HEMISPHERE

Star	Magnitudes	Separation"	Comments
ζ Aquarii	4.4, 4.6	2.0	Not easy. Slow binary
α Capricorni	3.8, 4.5	376.0	Naked-eye pair. Very easy
α Centauri	0.0, 1.7	22.0	Very easy. Binary, period 80 years
α Circini	3.4, 7.7	15.8	Easy
δ Corvi	3.0, 8.4	24.2	Very easy
α Crucis	1.4, 1.9	4.7	Superb pair
γ Crucis	1.6, 6.7	110.6	Optical pair, easy
θ Eridani	3.4, 4.4	8.5	Splendid easy pair
ε Hydrae	3.5, 6.9	2.9	Brighter star is a close binary
γ Leporis	3.8, 6.4	95.0	Very wide and easy
β Orionis	0.1, 7.0	9.2	Rigel. Faint optical companion
θ Orionis	6.0, 7.0, 7.5, 8.0	8.7, 12.9, 13.4, 19.2	Multiple star in the Great Nebula. Four stars visible with small telescopes
β Phoenicis	4.1, 4.1	1.3	Slow binary. Small separation
γ Puppis	5.0, 8.3	12.8	Slow binary
α Scorpii	0.9, 6.5	2.9	Antares. Red and green
β Scorpii	2.9, 5.1	13.7	Brighter star is a close double
ν Scorpii	4.3, 6.5	41.4	Both stars again double
β Tucanae	4.5, 4.9	27.1	Both double, but rather close
γ Velorum	2.2, 4.8	41.0	Wide and easy
δ Velorum	2.0, 6.5	2.9	Easy
γ Virginis	3.6, 3.6	below 5	Binary, 180 years. Closing, and becoming more difficult

VARIABLE STARS NORTHERN HEMISPHERE

Star	Magnitude min.–max.	Period days	Type	Comments
R Andromedae	5.9–14.9	397	Mira	Reddish
η Aquilae	3.7–4.7	7.2	Cepheid	Naked eye
R Arietis	7.5–13.7	187	Mira	Reddish
U Arietis	6.4–15.2	371	Mira	Reddish
ε Aurigae	3.3–4.2	9,899	Eclipsing	Next minimum due in 1982
γ Cassiopeiae	1.6–3.2	—	Irregular	Usually about 2.3
ρ Cassiopeiae	4.1–6.2	—	Irregular	Usually about 4.9
R Cassiopeiae	5.5–13.0	431	Mira	Reddish
δ Cephei	3.5–4.4	5.4	Cepheid	Naked eye
R Coronae Borealis	5.8–15.0	—	Irregular	Prototype star
χ Cygni	3.3–14.2	407	Mira	Naked eye at maximum
R Cygni	5.5–14.2	436	Mira	Reddish
SS Cygni	8.1–12.1	—	Irregular	Flares up about every 50 days
U Delphini	5.6–7.5	—	Irregular	Binocular
R Draconis	6.0–13.0	246	Mira	Reddish
ζ Geminorum	3.7–4.3	10.2	Cepheid	Naked eye
η Geminorum	3.1–3.9	±230	Semi-regular	Naked eye
α Herculis	3.0–4.0	±100	Semi-regular	Naked eye
R Leonis	5.4–10.5	313	Mira	Easy. Very red
β Lyrae	3.4–4.0	12.9	Eclipsing	Always in variation
X Ophiuchi	5.9–9.2	334	Mira	Easy, reddish
α Orionis	0.0–0.9	±2,070	Semi-regular	Always brilliant
β Pegasi	2.4–2.8	±36	Semi-regular	Naked eye
β Persei	2.1–3.3	2.9	Algol	Prototype star
ρ Persei	3.3–4.2	33–55	Semi-regular	Naked eye. Reddish
λ Tauri	3.3–4.2	3.95	Algol	Naked eye
Z U. Majoris	6.6–9.1	198	Semi-regular	Easy
R Virginis	6.2–12.1	146	Mira	Reddish

VARIABLE STARS SOUTHERN HEMISPHERE

Star	Magnitude min.–max.	Period days	Type	Comments
R Arae	5.9–6.9	4.4	Algol	Easy, binocular
η Carinae	−0.8–7.9	—	Irregular	Unique object
R Carinae	3.9–10.0	381	Mira	Easy, naked eye
R Centauri	5.4–11.8	547	Mira	Reddish
o Ceti	1.7–10.1	331	Mira	Naked eye at maximum. Prototype of its class
S Crucis	6.6–7.7	4.7	Cepheid	Binocular object
β Doradûs	4.5–5.7	9.8	Cepheid	Naked eye
R Hydrae	4.0–10.0	386	Mira	Naked eye near maximum
R Leporis	5.9–10.5	432	Mira	Very red. Easy
δ Librae	4.8–6.1	2.3	Algol	Naked eye
κ Pavonis	4.0–5.5	9.1	W Virginis	Naked eye
Y Pavonis	5.7–8.5	233	Semi-regular	Period fluctuates
ζ Phoenicis	3.5–4.1	1.7	Eclipsing	Naked eye
R Pictoris	6.7–10.0	171	Mira	Reddish
L² Puppis	3.4–6.2	141	Semi-regular	Naked eye
V Puppis	4.3–5.1	1.45	Beta Lyrae	Naked eye
X Sagittarii	4.1–5.1	7.0	Cepheid	Naked eye
R Scuti	5.7–8.6	±144	RV Tauri	Binocular
RR Telescopii	6.5–16.5	—	Novalike	Occasional outbursts

Stellar types 2

BRIGHTEST STARS NORTHERN HEMISPHERE

Star	Constellation	Magnitude	Distance (light-years)	Luminosity (Sun = 1)
Arcturus	Boötes	−0.1	36	115
Vega	Lyra	0.0	27	55
Capella	Auriga	0.0	45	150
Procyon	Canis Minor	0.4	11	7
Betelgeuse	Orion	var	520	15,000
Altair	Aquila	0.8	16	10
Aldebaran	Taurus	0.8	68	165
Pollux	Gemini	1.2	35	34
Deneb	Cygnus	1.3	1,600	60,000
Regulus	Leo	1.4	84	170

BRIGHTEST STARS SOUTHERN HEMISPHERE

Star	Constellation	Magnitude	Distance (light-years)	Luminosity (Sun = 1)
Sirius	Canis Major	−1.4	8.7	26
Canopus	Carina	−0.7	650(?)	80,000(?)
Alpha Centauri	Centaurus	−0.3	4.3	1.5
Rigel	Orion	0.1	900	60,000
Achernar	Eridanus	0.5	118	720
Agena	Centaurus	0.6	490	10,500
Acrux	Crux	0.8	370	3,200 2,000 (binary)
Antares	Scorpio	0.8	520	9,600
Spica	Virgo	0.9	220	1,800
Fomalhaut	Piscis Austrinus	1.2	23	15
Beta Crucis	Crux	1.3	490	6,000

CLUSTERS, NEBULAE AND GALAXIES NORTHERN HEMISPHERE

Constellation	Catalogue number	Type	Comments
Andromeda	M31	Great Spiral	Naked eye. Two companion galaxies, M32 and NGC 205, are easy objects
Auriga	M38	Open cluster	Easy in binoculars
Cancer	M44	Open cluster	Praesepe. Naked eye. Best seen with a low magnification
Cancer	M67	Open cluster	Naked eye with some difficulty. Easy in binoculars
Canes Venatici	M3	Globular	Visible in binoculars
Canes Venatici	M51	Galaxy	The Whirlpool Spiral. Faint in small telescopes
Cassiopeia	H VI 31	Open cluster	Binocular
Gemini	M35	Open cluster	Naked eye, splendid sight in a small telescope
Hercules	M13	Globular	Just visible with naked eye. Best northern globular
Hercules	M92	Globular	Little inferior to M13
Lyra	M57	Planetary	Ring Nebula. Easy with moderate telescopes, between Beta and Gamma Lyrae
Pegasus	M15	Globular	Magnitude 6. Bright, condensed
Perseus	NGC 869 NGC 884	Twin clusters	Naked eye. Lovely in binoculars or a small telescope
Serpens	M5	Globular	Just below magnitude 6.

Taurus	M45	Open cluster	The Pleiades. Naked eye
Taurus	Mel 25	Open cluster	The Hyades, around Aldebaran. Very scattered
Triangulum	M33	Galaxy	Spiral. Has been reported with the naked eye but can be very elusive telescopically
Ursa Major	M81	Galaxy	Not difficult. M82 lies near by
Ursa Major	M97	Planetary	Owl Nebula. Very faint in small telescopes
Virgo	M87	Galaxy	Giant elliptical, radio source, but very faint in small telescopes
Vulpecula	M27	Planetary	Dumb-bell Nebula. Not difficult

CLUSTERS, NEBULAE AND GALAXIES SOUTHERN HEMISPHERE

Constellation	Catalogue number	Type	Comments
Aquarius	NGC 7293	Planetary nebula	Magnitude 6.5. Easy object
Ara	NGC 6193	Open cluster	Easy in binoculars
Ara	NGC 6362	Globular	Easy
Canis Major	M41	Open cluster	Naked eye
Carina	NGC 2156	Open cluster	Naked eye. Rich
Carina	NGC 3372	Nebula	Keyhole Nebula, around Eta Carinae. Superb spectacle
Centaurus	NGC 5139	Globular	Omega Centauri, finest of all globulars
Centaurus	NGC 3766	Open cluster	Easy in binoculars
Cetus	M77	Galaxy	Seyfert galaxy. Magnitude 8.9 rather faint in small telescopes
Crux	NGC 4755	Open cluster	The Jewel Box, around Kappa Crucis. Very prominent
Hydra	M48	Open cluster	Binocular
Orion	M42	Nebula	Great Nebula. Naked eye
Puppis	M47	Open cluster	Naked eye
Sagittarius	M8	Nebula	Lagoon Nebula. Not difficult
Sagittarius	M20	Nebula	Trifid Nebula. Very faint in small telescopes
Sagittarius	M22	Globular	Binocular
Sagittarius	M23	Open cluster	Binocular
Sagittarius	M24	Open cluster	Binocular
Scorpio	M4	Globular	Easy
Scorpio	M6	Open cluster	Naked eye. Superb cluster
Scorpio	M7	Open cluster	Glorious, brilliant naked-eye cluster
Scorpio	M80	Globular	Easy, although fainter than M4
Scutum	M11	Open cluster	Wild Duck, lovely cluster
Tucana	NGC 104	Globular	47 Tucanae. Inferior only to Omega Centauri
Vela	NGC 2547	Open cluster	Naked eye

GREEK ALPHABET

α alpha	η eta	ν nu	τ tau
β beta	θ theta	ξ xi	υ upsilon
γ gamma	ι iota	o omicron	φ phi
δ delta	κ kappa	π pi	χ chi
ε epsilon	λ lambda	ρ rho	ψ psi
ζ zeta	μ mu	σ sigma	ω omega

Star-chart index

138

Constellation	Genitive	Popular name	Map number
Hydra	Hydrae	Water Serpent	1, 3, 7, 11, 12, 13, 14, 16
Hydrus	Hydri	Water Serpent	10, 12, 14, 16
Indus	Indi	Indian	3, 5, 10, 12, 16
Lepus	Leporis	Hare	1, 7, 11, 14
Libra	Librae	Scales	1, 3, 10, 14, 15
Lupus	Lupi	Wolf	3, 10, 12, 14, 16
Mensa	Mensae	Table Mountain	Not shown
Microscopium	Microscopii	Microscope	Not shown
Monoceros	Monocerotis	Unicorn	1, 7, 11, 13
Musca	Muscae	Fly	10, 12, 14, 16
Norma	Normae	Square, Level	3, 10, 14, 16
Octans	Octantis	Octant	10, 12, 14, 16
Ophiuchus	Ophiuchi	Serpent Holder	3, 5, 9, 10, 14, 15
Pavo	Pavonis	Peacock	10, 12, 14, 16
Phoenix	Phoenicis	Phoenix	5, 10, 12, 14, 16
Pictor	Pictoris	Painter's Easel	7, 10, 12, 14, 16
Piscis Austrinus	Piscis Austrini	Southern Fish	3, 5, 10, 12, 16
Puppis	Puppis	Stern	1, 7, 12, 14
Pyxis	Pyxidis	Mariner's Compass	1, 7, 12, 14
Reticulum	Reticuli	Net	10, 12, 14, 16
Sagittarius	Sagittarii	Archer	3, 5, 10, 14, 16
Scorpius	Scorpii	Scorpion	3, 10, 14, 16
Sculptor	Sculptoris	Sculptor	5, 10, 12, 16
Sextans	Sextantis	Sextant	1, 7, 11, 13
Telescopium	Telescopii	Telescope	3, 10, 14, 16
Triangulum Australe	Trianguli Australis	Southern Triangle	10, 12, 14, 16
Tucana	Tucanae	Toucan	10, 12, 14, 16
Vela	Velorum	Sail	1, 12, 14, 16
Virgo	Virginis	Virgin	1, 3, 8, 13, 15
Volans	Volantis	Flying Fish	10, 12, 14, 16

MAP REFERENCES

The tables below indicate the appropriate star map for any hour and month of the year. Read across and downwards to locate the required map, which appears between pages 78 and 93.

NORTHERN HEMISPHERE

UT	18.00	20.00	22.00	Midnight	02.00	04.00	06.00
Jan.	5–6	7–8	7–8	7–8	1–2	1–2	1–2
Feb.	7–8	7–8	7–8	1–2	1–2	1–2	3–4
Mar.	7–8	7–8	1–2	1–2	1–2	3–4	3–4
Apr.		1–2	1–2	1–2	3–4	3–4	
May		1–2	1–2	3–4	3–4	3–4	
Jun.			3–4	3–4	3–4		
Jul.		3–4	3–4	3–4	5–6	5–6	
Aug.		3–4	3–4	5–6	5–6	5–6	
Sep.	3–4	3–4	5–6	5–6	5–6	7–8	7–8
Oct.	3–4	5–6	5–6	5–6	7–8	7–8	7–8
Nov.	5–6	5–6	5–6	7–8	7–8	7–8	1–2
Dec.	5–6	5–6	7–8	7–8	7–8	1–2	1–2

SOUTHERN HEMISPHERE

UT	18.00	20.00	22.00	Midnight	02.00	04.00	06.00
Jul.		13–14	15–16	15–16	15–16	9–10	9–10	9–10
Aug.	15–16	15–16	15–16	9–10	9–10	9–10	9–10
Sep.	15–16	15–16	9–10	9–10	9–10	11–12	11–12
Oct.		9–10	9–10	9–10	11–12	11–12	
Nov.		9–10	9–10	11–12	11–12	11–12	
Dec.			11–12	11–12	11–12		
Jan.		11–12	11–12	11–12	13–14	13–14	
Feb.		11–12	11–12	13–14	13–14	13–14	
Mar.	11–12	11–12	13–14	13–14	13–14	15–16	15–16
Apr.	11–12	13–14	13–14	13–14	15–16	15–16	15–16
May	13–14	13–14	13–14	15–16	15–16	15–16	9–10
Jun.	13–14	13–14	15–16	15–16	15–16	9–10	9–10

Moon-map index

141

General index

142

143